T0181713

Indian Statistical Institute Series

The *Indian Statistical Institute Series* publishes high-quality content in the domain of mathematical sciences, bio-mathematics, financial mathematics, pure and applied mathematics, operations research, applied statistics and computer science and applications with primary focus on mathematics and statistics. Editorial board comprises of active researchers from major centres of Indian Statistical Institute. Launched at the 125th birth Anniversary of P. C. Mahalanobis, the series will publish textbooks, monographs, lecture notes and contributed volumes. Literature in this series will appeal to a wide audience of students, researchers, educators, and professionals across mathematics, statistics and computer science disciplines.

More information about this series at http://www.springer.com/series/15910

N. S. Narasimha Sastry · Manoj Kumar Yadav
Editors

Group Theory and Computation

Springer

Editors
N. S. Narasimha Sastry
Department of Mathematics
Indian Institute of Technology Dharwad
Dharwad, Karnataka, India

Manoj Kumar Yadav
Harish-Chandra Research Institute
Allahabad, Uttar Pradesh, India

ISSN 2523-3114 ISSN 2523-3122 (electronic)
Indian Statistical Institute Series
ISBN 978-981-13-4724-5 ISBN 978-981-13-2047-7 (eBook)
https://doi.org/10.1007/978-981-13-2047-7

Cover photo: Reprography & Photography Unit, Indian Statistical Institute, Kolkata, India

This Springer imprint is published by the registered company Springer Nature Singapore Pte Ltd.
The registered company address is: 152 Beach Road, #21-01/04 Gateway East, Singapore 189721, Singapore

Preface

This volume is a collection of invited articles by some of the leading and very active researchers in the theory of finite groups and their representations, and the Monster group, with an emphasis on computational aspects. Among the authors are participants who led workshops and delivered invited talks in the international program "Group Theory and Computational Methods" held at the International Center for Theoretical Sciences (ICTS), Bangalore, from November 05–14, 2016. The program comprised of two parts: workshops (November 05–09, 2016) and discussion meetings (November 11–14, 2016).

The workshops comprised of the following five minicourses of 6-h duration each:

(1) Computational homological algebra
(2) Computational representation theory
(3) Computational aspects of finite p-groups
(4) Computer algebra system GAP
(5) Introduction to Monster simple group and Moonshine

The discussion meetings comprised of 20 invited talks by eminent mathematicians. They presented recent developments and problems of current mathematical interest on a variety of topics in group theory and related areas. Poster presentation session consisted of 11 posters by young researchers.

The topics of the articles include finite loops, non-abelian tensor product, periodic groups, character table of finite groups, computing subgroups using computer algebra system GAP, Majorana theory related to the Monster group, groups with abelian automorphism groups, unit groups of integral group rings, and Camina groups and generalizations.

We thank the authors for writing the articles and our colleagues for carefully refereeing them. We also thank the speakers and the participants of the workshops and discussion meetings mentioned above whose participation helped closer mathematical scrutiny of the themes discussed.

We thank ICTS, Bangalore, for the excellent facilities and support for the smooth and successful conduct of the program. We also thank Springer for publishing this volume, and Shamim Ahmad, for his friendly and efficient handling of the publication process.

Dharwad, India N. S. Narasimha Sastry
Allahabad, India Manoj Kumar Yadav

Contents

Editors and Contributors

About the Editors

N. S. Narasimha Sastry is currently a visiting professor at the Indian Institute of Technology Dharwad, Karnataka, India. Prior to this, he was a professor of mathematics and head of the Indian Statistical Institute, Bangalore Centre, India. He received his Ph.D. from the University of Pittsburgh, Pennsylvania, USA, and has held visiting positions at the Tata Institute of Fundamental Research, Mumbai, India; Michigan State University, East Lansing, USA; The University of Western Australia, Perth, Australia; Rutgers University–New Brunswick, USA; The University of Florida, Gainesville, USA; and Ghent University, Belgium. In addition to publishing more than 30 research articles in various journals, he has also edited books: *Buildings, Finite Geometries and Groups* and *Groups of Exceptional Type, Coxeter Groups and Related Geometries*, both published by Springer, as well as *Essays in Geometric Group Theory*, published by the Ramanujan Mathematical Society. He also co-edited *Perspectives in Mathematical Sciences* (Volumes 1 and 2), published by World Scientific. His research interests are finite groups including finite simple groups, geometries related to finite simple groups, algebraic codes, Coxeter groups, and the Monster group.

Manoj Kumar Yadav is Professor of Mathematics at the Harish-Chandra Research Institute, Allahabad, India. He is also associated with the Homi Bhabha National Institute, Mumbai, India. He received his PhD in Mathematics from Kurukshetra University, Kurukshetra (2002). Professor Yadav has been awarded the Indian National Science Academy Medal for Young Scientists (2009) and the Department of Science & Technology, Science and Engineering Research Council (SERC), fellowship Fast Track Scheme for Young Scientists (2005). He is a member of The National Academy of Sciences, India (NASI). His research interests lie in group theory, particularly automorphisms, conjugacy classes and Schur multipliers of groups. He has published over 25 research papers in various respected journals, conference proceedings and contributed volumes.

Contributors

Katharina Artic Lehrstuhl B für Mathematik, RWTH Aachen University, Aachen, Germany

Valeriy G. Bardakov Sobolev Institute of Mathematics, Novosibirsk, Russia; Novosibirsk State University, Novosibirsk, Russia; Novosibirsk State Agrarian University, Novosibirsk, Russia

Silvio Dolfi Dipartimento di Matematica U. Dini, Università degli Studi di Firenze, Firenze, Italy

Marcel Herzog School of Mathematical Sciences, Tel-Aviv University, Ramat-Aviv, Tel-Aviv, Israel

Gerhard Hiss Lehrstuhl D für Mathematik, RWTH Aachen University, Aachen, Germany

Alexander Hulpke Department of Mathematics, Colorado State University, Fort Collins, CO, USA

Alexander A. Ivanov Department of Mathematics, Imperial College, London, UK

Rahul Dattatraya Kitture School of Mathematics, Harish-Chandra Research Institute, Jhunsi, Allahabad, India; Homi Bhabha National Institute, Mumbai, India

Mark L. Lewis Department of Mathematical Sciences, Kent State University, Kent, OH, USA

Patrizia Longobardi Dipartimento di Matematica, Università di Salerno, Fisciano (Salerno), Italy

Sugandha Maheshwary Indian Institute of Science Education and Research, Mohali, Mohali, Punjab, India

Mercede Maj Dipartimento di Matematica, Università di Salerno, Fisciano (Salerno), Italy

Gabriel Navarro Departament of Mathematics, Universitat de València, Burjassot, Valencia, Spain

Mikhail V. Neshchadim Sobolev Institute of Mathematics and Novosibirsk State University, Novosibirsk, Russia

Emanuele Pacifici Dipartimento di Matematica F. Enriques, Università degli Studi di Milano, Milano, Italy

Inder Bir S. Passi Indian Institute of Science Education and Research, Mohali, Mohali, Punjab, India; Centre for Advanced Study in Mathematics, Panjab University, Chandigarh, India

Lucia Sanus Departament de Matemàtiques, Facultat de Matemàtiques, Universitat de València, Valencia, Spain

Manoj K. Yadav School of Mathematics, Harish-Chandra Research Institute, Jhunsi, Allahabad, India; Homi Bhabha National Institute, Mumbai, India

On Right Conjugacy Closed Loops of Twice Prime Order

Katharina Artic and Gerhard Hiss

2010 Mathematics Subject Classification 20N05 · 20B10 · 20E45

1 Introduction

A *quasigroup* $\mathcal{L}$ is a set with a binary operation $* : \mathcal{L} \times \mathcal{L} \to \mathcal{L}$, such that every equation $x * a = b$ or $a * x = b$ with $a, b \in \mathcal{L}$ has a unique solution x. In this case, for every $a \in \mathcal{L}$, the *right multiplication* $R_a : \mathcal{L} \to \mathcal{L}, x \mapsto x * a$ is a permutation of $\mathcal{L}$ (and of course so is every *left multiplication*). A quasigroup is a *loop*, if it contains an identity element. Thus, a group is just a loop, in which the operation is associative, and we will indeed view groups as loops.

In the following, we will only consider finite loops. Let $\mathcal{L}$ be a (finite) loop, whose identity element we denote by e. The *right multiplication group of* $\mathcal{L}$ is the group $G := \langle R_a \mid a \in \mathcal{L} \rangle$, a subgroup of the symmetric group on $\mathcal{L}$. Clearly, G acts faithfully and transitively on $\mathcal{L}$ and R_e is the identity element of G, which we denote by 1. Let $H \leq G$ denote the stabilizer in G of $e \in \mathcal{L}$, and let $T := \{R_a \mid a \in \mathcal{L}\}$. Then, T is a transversal for $H^g \backslash G := \{H^g x \mid x \in G\}$ for every $g \in G$, the identity element of G is contained in T, and $\langle T \rangle = G$. The triple (G, H, T) is called the *envelope* of $\mathcal{L}$, a group theoretic object..

Conversely, starting from group theory, one defines a *loop folder* to be a triple (G, H, T) of a finite group G, a subgroup $H \leq G$ and a subset $T \subseteq G$ with $1 \in T$, such that T is a transversal for $H^g \backslash G$ for every $g \in G$. Given a loop folder (G, H, T), one can construct a loop $(\mathcal{L}, *)$ on the set $H \backslash G$ of right cosets of H in G. However, the envelope of $\mathcal{L}$ need not be equal to (G, H, T). In contrast to the right multiplication

K. Artic
Lehrstuhl B für Mathematik, RWTH Aachen University, 52056 Aachen, Germany
e-mail: katharina.artic@math.rwth-aachen.de

G. Hiss (✉)
Lehrstuhl D für Mathematik, RWTH Aachen University, 52056 Aachen, Germany
e-mail: gerhard.hiss@math.rwth-aachen.de

N. S. N. Sastry and M. K. Yadav (eds.), *Group Theory and Computation*,
Indian Statistical Institute Series, https://doi.org/10.1007/978-981-13-2047-7_1

group of $\mathcal{L}$, in general, the group G will not act faithfully on $\mathcal{L}$, and the transversal T will not generate G. On the other hand, it is not difficult to construct the envelope of $\mathcal{L}$ from (G, H, T).

These results, as well as the notion of *loop folder* and *envelope of a loop* are contained in [2, Section 1]. However, the connection between loops and their envelopes goes back to Baer [3].

Let $\mathcal{L}$ be a loop with envelop (G, H, T). We say that $\mathcal{L}$ is *right conjugacy closed*, or an RCC loop, if $T = \{R_a \mid a \in \mathcal{L}\}$ is closed under conjugation by itself. Clearly, this is the case if and only if T is invariant under conjugation in $G = \langle T \rangle$; in other words, if T is a union of conjugacy classes of G. We shortly say that T is G-invariant in the following. Thus, an RCC loop gives rise to a G-invariant transversal of H, the stabilizer of e in G. (A G-invariant transversal of a subgroup H of a group G is sometimes called a *distinguished transversal* in the literature.) On the group-theoretic side, this leads to the notion of an RCC loop folder. This is a loop folder (G, H, T), where T is G-invariant. More definitions regarding loop folders are given at the beginning of Sect. 3.

It has been shown by Drápal [9] that an RCC loop of prime order is a group. In this paper, we determine all RCC loops of order $2p$, where p is an odd prime. In order to achieve this, we first describe the possible envelopes (G, H, T) of such loops. Our approach is group theoretic. In Sects. 2 and 3, we show that if (G, H, T) is an RCC loop folder such that G acts faithfully on $H\backslash G$ and the index $|G\colon H|$ is the product of two distinct primes, then G acts imprimitively on $H\backslash G$ (Theorem 3.1). This result uses the classification of the finite simple groups and is based on the classification of finite primitive permutation groups of square-free degree by Li and Seress, and on the determination of the minimal degrees of permutation representations of finite groups of Lie type by Patton, Cooperstein and Vasilyev. For the purpose of our further investigation, it would suffice to enumerate the primitive permutation groups of degree $2p$ for odd primes p; we are not aware of any result in this direction which does not rely on the classification of the finite simple groups.

In Sect. 4, we continue with some basic results on permutation groups of degree p and give a new proof of Drápal's theorem on RCC loops of prime order (Corollary 4.2).

Let (G, H, T) be the envelope of an RCC loop of order $2p$, where p is an odd prime. Using Theorem 3.1 mentioned above, we may now assume that there is a subgroup $K \lneqq G$ with $H \lneqq K$, and also that one of the indices $|G\colon K|$ or $|K\colon H|$ is equal to 2, and the other index is equal to p. This configuration is analysed in Sect. 5 with elementary group theoretical methods. It turns out that there are three possible types for G. First, G can be isomorphic to the wreath product $C_p \wr C_2$, where C_p and C_2 denote (cyclic) groups of order p and 2, respectively. Second, G can be isomorphic to a subgroup of $\mathrm{Aff}(1, p)$, the affine group over $\mathbb{F}_p$. Third, G can be isomorphic to a group $K \times \langle a \rangle$, where K is an odd order subgroup of $\mathrm{Aff}(1, p)$ and a is an element of order 2 (Theorem 5.13). In particular, G is soluble. Ultimately, our results rely on the classification of the finite simple groups. One could avoid this by assuming from the outset that G is soluble. This would lead to exactly the same list of RCC loops of order $2p$, but of course without the guarantee to have found them all.

In Sect. 6, we determine the number of isomorphism classes of loops of order $2p$ (Theorem 6.5).

Finally, Sect. 7 introduces a series of examples of RCC loops of order $q^2 - 1$ and multiplication groups $\mathrm{GL}(2, q)$ (Proposition 7.1). For $q = 4$, we obtain a loop of order $3 \cdot 5$, whose multiplication group is not soluble. These examples indicate that a generalization of our results to RCC loops of order pq for distinct primes p and q could be substantially more difficult.

This is a good place to discuss some related results. In [18, Theorem A], Stein shows that if T is a conjugacy class in a finite group and at the same time, a tranversal to a subgroup, then $\langle T \rangle$ is soluble. This result uses the classification of the finite simple groups. Without the classification, but with the help of the Odd Order Theorem, Csörgő and Niemenmaa in [5] obtain the solubility of the full multiplication group of a loop under certain conditions on the stabilizer of a point. Their paper contains further references for results along this line. In [6], Csörgő and Drápal characterize left conjugacy closed loops inside the class of nilpotent loops of nilpotency class two. In the same paper, these authors also determine the nilpotent left conjugacy closed loops of order p^2 for primes p. In [14, Theorem 4.15], Kunen shows that for each odd prime p there is exactly one non-associative conjugacy closed loop of order $2p$, up to isomorphism (a loop is conjugacy closed, if it is both left and right conjugacy closed). Burn shows in [4] that every Bol loop of order p^2 or $2p$ for a prime p is a group. Finally, in [7, Theorem 7.1], Daly and Vojtěchovský determine the number of nilpotent loops of order $2p$, where again p is a prime, up to isomorphism.

This paper builds upon the PhD thesis of the first author [1], written under the direction of the second author and Alice Niemeyer. Theorem 3.1 is contained in this thesis, but also a complete classification of all RCC loops of order at most 30. These have been incorporated into the GAP package *Loops* of Nagy and Vojtěchovský [16]. The classification of the RCC loops of order $2p$ is, to the best of our knowledge, new. The examples computed in [1] were of considerable importance for confirming our theoretical results of Sect. 6. The example of an RCC loop of order 15 and multiplication group $\mathrm{GL}(2, 4)$ contained in [1], gave rise to the series of examples constructed in Sect. 7.

Our group theoretical notation is standard. For example, we write G' for the commutator subgroup of the group G. We do recall the notion of an almost simple group and that of the core of a subgroup in the introductions to Sects. 2 and 3, respectively. As already indicated above, a cyclic group of order n is denoted by C_n, and the symmetric and alternating groups of degree n are denoted by S_n and A_n, respectively.

2 Primitive Permutation Groups of Square-free Degree

We begin with a remark on the sizes of conjugacy classes in almost simple groups. Recall that a group G is *almost simple*, if there is a non-abelian finite simple group S such that $S \leq G \leq \mathrm{Aut}(S)$ (where S is identified with the group of inner automorphisms of S). In this context, S is called the *socle* of G.

Remark 2.1 Let G be an almost simple group with socle S. Denoted by l the smallest index of any proper subgroup of S. Since S is simple, l is a lower bound for the size of those non-trivial conjugacy classes of G lying in S. Let $g \in G \setminus S$. Then, we have

$$|G\colon C_G(g)| = \frac{|G|}{|SC_G(g)|} \cdot \frac{|S|}{|S \cap C_G(g)|}$$
$$= \frac{|G|}{|SC_G(g)|} \cdot \frac{|S|}{|C_S(g)|}.$$

Notice that, if $C_S(g) = S$, the element g acts trivially on S which implies that $g = 1$. Hence, we have $C_S(g) \lneq S$. Thus, l is a lower bound on the size of all non-trivial conjugacy classes of G.

The following theorem combines some major results by Li and Seress on finite primitive permutation groups of square-free degree, and by Patton, Cooperstein and Vasilyev on the minimal degrees of permutation representations of finite groups of Lie type.

Theorem 2.2 *Let G be a finite primitive permutation group of degree n (i.e. G acts faithfully and primitively on a set of n points). Suppose that n is square free (i.e. $p^2 \nmid n$ for all primes p). Then every non-trivial conjugacy class of G has at least n elements, or one of the following holds:*

(a) *We have $n = p$ is a prime and G is isomorphic to a subgroup of* $\mathrm{Aff}(1, p)$,
(b) *We either have have $G = S_8$ and $n \in \{35, 105\}$, or $G = J_1$ and $n = 2926$, or $G = \mathrm{PGL}(2, r)$ for an odd prime r and $n = r(r + 1)/2$,*
(c) *or G is almost simple and* $\mathrm{soc}(G)$ *and n occur in* Table 1. *There, r denotes a prime power.*

Proof By [15, Theorem 1] we either have that n is a prime and $G \leq \mathrm{Aff}(1, n)$ as in Case (a), or G is almost simple and $S := \mathrm{soc}(G)$ as well as n appear in the paper [15] by Li and Seress.

The cases when S is isomorphic to an alternating group, are listed in [15, Table 1]. If S is as in [15, Table 1, Line 1], then $S = A_c$ and $n = \binom{c}{k}$ with $1 \leq k \leq c - 1$. For reasons of symmetry it suffices to consider the case $k \leq c/2$. Table 2 lists the size $s(G)$ of the smallest non-trivial conjugacy class of G for all G with $S \in \{A_5, A_6, A_7, A_8\}$. This table, easily compiled or verified with GAP [20], proves our claim for $5 \leq c \leq 8$. If $c \geq 9$, by [8, Theorems 5.2A,B], the subgroups of A_c or S_c which have an index less then $c(c - 1)/2$ do not occur as centralizers of non-trivial elements. Hence, the

Table 1 Primitive groups of degree n which might have a non-trivial conjugacy class of length less than n

	soc(G)	n	Restrictions
(i)	A_c	$\binom{c}{k}$	$3 \leq k \leq c-3$
(ii)	A_{2a}	$\frac{1}{2}\binom{2a}{a}$	$a \in \{6, 9, 10, 12, 36\}$
(iii)	PSL(m, r)	$\frac{\prod_{i=0}^{k-1}(r^{m-i}-1)}{\prod_{i=1}^{k}(r^i-1)}$	$2 \leq k \leq m-2$, $(m, r) \notin \{(4, 2), (5, 2)\}$
(iv)	PSL(m, r)	$\frac{\prod_{i=0}^{2k-1}(r^{m-i}-1)}{(\prod_{i=1}^{k}(r^i-1))^2}$	$1 \leq k < m/2$, $m \geq 3$, $(m, r) \neq (3, 2)$
(v)	PSL$(2, r)$	$\sqrt{r}(r+1)/2$	$\sqrt{r}$ an odd prime, soc$(G) < G$, $r > 9$
(vi)	PSL$(2, r)$	$r(r^2-1)/24$	r a prime, $r \equiv \pm 3 \pmod 8$, $r \notin \{5, 11\}$
(vii)	PSL$(2, r)$	$r(r^2-1)/48$	r a prime, $r \equiv \pm 1 \pmod 8$, $r \notin \{7, 17, 23\}$
(viii)	PSL$(2, r)$	$r(r^2-1)/120$	r a prime, $r \equiv \pm 1 \pmod{10}$, $r \notin \{11, 19, 29, 31, 41, 59\}$
(ix)	PSU$(4, r)$	$(r^2+1)(r^3+1)$	
(x)	PSp$(2m, 2)$	$4^m - 1$	$m \geq 3$
(xi)	PSp$(2m, r)$	$\frac{(r^{2m}-1)(r^{2m-2}-1)}{(r^2-1)(r-1)}$	$m \geq 3$
(xii)	$\Omega(2m+1, r)$	$\frac{(r^{2m}-1)(r^{2m-2}-1)}{(r^2-1)(r-1)}$	$m \geq 3$
(xiii)	P$\Omega^-(2m, r)$	$\frac{(r^m+1)(r^{2m-2}-1)(r^{m-2}-1)}{(r^2-1)(r-1)}$	$m \geq 3$, r even
(xiv)	P$\Omega^-(2m, r)$	$\frac{(r^m+1)(r^{2m-2}-1)(r^{2m-4}-1)(r^{m-3}-1)}{(r^3-1)(r^2-1)(r-1)}$	$m \equiv 0 \pmod 4$, r even
(xv)	P$\Omega^+(2m, 2)$	$(2^m-1)(2^{m-1}+1)$	$m \geq 5$ odd
(xvi)	P$\Omega^+(2m, r)$	$\frac{(r^m-1)(r^{2m-2}-1)(r^{m-2}+1)}{(r^2-1)(r-1)}$	$m \geq 3$, r even
(xvii)	P$\Omega^+(2m, r)$	$\frac{(r^m-1)(r^{2m-2}-1)(r^{2m-4}-1)(r^{m-3}+1)}{(r^3-1)(r^2-1)(r-1)}$	$m \equiv 3 \pmod 4$, r even
(xviii)	$E_7(r)$	$\frac{(r^{18}-1)(r^{14}-1)(r^4-r^2+1)}{(r^2-1)(r-1)}$	

Table 2 Smallest size of non-trivial conjugacy classes

G	A_5	S_5	A_6	S_6	$A_6.2_2$	$A_6.2_3$	Aut(A_6)	A_7	S_7	A_8	S_8
s	12	15	40	15	36	45	30	70	21	105	28

non-trivial conjugacy classes of A_c or S_c have at least $c(c-1)/2$ elements, proving our claim for $k \leq 2$. The case $3 \leq k \leq c-3$ appears as Case (c)(i) in our statement.

In the remaining cases of [15, Table 1], a look at Table 2 shows that all non-trivial conjugacy classes of G have at least n elements except for

- $G = S_8$ and $n \in \{35, 105\}$,
- $S = A_{2a}$ and $n = \binom{2a}{a}/2$ with $a \in \{6, 9, 10, 12, 36\}$.

These cases appear as Case (b) and Case (c)(ii), respectively, in our statement.

The cases when S is a sporadic simple group are listed in [15, Table 2]. Using GAP, we only find the one exception listed in Case (b).

In [15, Table 3], the case where S is a classical group are considered. For some small parameter values, we have verified our claim directly with GAP. These cases are listed in the column headed *Restrictions* of Table 1, and are not commented on any further below. In the following, we refer to the line numbers of [15, Table 3]. Suppose that S is as in Line 1. Then, $S = \mathrm{PSL}(m, r)$ and

$$n = \prod_{i=0}^{k-1}(r^{m-i} - 1)/\prod_{i=1}^{k}(r^i - 1)$$

with $1 \leq k < m$. For $k = 1$ or $k = m-1$, we have $n = (r^m - 1)/(r-1)$. If $(m, r) \in \{(2, 5), (2, 7), (2, 9), (2, 11), (4, 2)\}$, a computation with GAP shows that the non-trivial conjugacy classes of G have more than n elements. Otherwise, n is the smallest index of any proper subgroup of S by [13, Table 5.2.A]. Applying Remark 2.1, we see that the non-trivial conjugacy classes of G have at least n elements. The case $2 \leq k \leq m-2$ is listed as Case (c)(iii) in our statement.

The case when S is as in Line 2, is listed as Case (c)(iv) in our statement.

Suppose that S is as in Line 3 or 4. Then, $S = \mathrm{PSL}(2, r)$ and $n = r(r \pm 1)/2$. Since n is square free, we have $r = 4$ and $n \in \{6, 10\}$ or r is an odd prime. If $r = 4$, we have $S \cong A_5$, and Table 2 proves our claim. If r is an odd prime, then $\mathrm{Aut}(\mathrm{PSL}(2, r)) = \mathrm{PGL}(2, r)$ and hence $G = \mathrm{PSL}(2, r)$ or $G = \mathrm{PGL}(2, r)$. The conjugacy classes of these groups are well known. We find that only if $G = \mathrm{PGL}(2, r)$ and $n = r(r+1)/2$, there are non-trivial conjugacy classes of G with less than n elements. This case appears in Case (b) in our statement.

Suppose that S is as in Line 5. Then, $S = \mathrm{PSL}(2, r)$ and $n = \sqrt{r}(r+1)/2$. Since n is square free, r is the square of a prime number. The non-trivial conjugacy classes of $\mathrm{PSL}(2, r)$ have at least n elements. Hence $S \lneq G$. This case is listed as Case (c)(v) in our statement.

Suppose that S is as in one of the Lines 6, 7 or 8. Then $S = \mathrm{PSL}(2, r)$ and $n = r(r^2 - 1)/d$ with $d \in \{24, 48, 120\}$, and $r \equiv \pm 3 \pmod 8$ if $d = 24$, respectively $r \equiv \pm 1 \pmod 8$ if $d = 48$, respectively $r \equiv \pm 1 \pmod{10}$ if $d = 120$. In particular, r is odd. Since n is square free, $r = 9$ or r is an odd prime. If $r = 9$ we have $S \cong A_6$, and Table 2 proves our claim. The cases where r is an odd prime, are listed as Case (c)(vi) through Case (c)(viii) in our statement.

Suppose that S is as in Line 9. Then, $S = \mathrm{PSU}(m, r)$ with

$$n = \frac{(r^m - (-1)^m)(r^{m-1} - (-1)^{m-1})}{r^2 - 1}.$$

If $m = 2$, we have $S \cong \mathrm{PSL}(2, r)$ (see [19, Theorem 10.9]) and $n = r + 1$, a case we have already considered above. For $m = 3$ and $r = 5$, our claim can be verified with GAP. The case of $m = 4$ is listed as Case (c)(ix) in our statement. If $6 \mid m$ and $r = 2$, then $n = (2^m - 1)(2^{m-1} + 1)/3$ is not square free, as $2^6 - 1$ divides $2^m - 1$ and 3 divides $2^{m-1} + 1$. In the remaining cases, n is the smallest index of any proper subgroup of S (see [13, Table 5.2.A]). Thus, by Remark 2.1, the non-trivial conjugacy classes of G have at least n elements.

Suppose that S is as in Line 10. Then, $S = \mathrm{PSp}(2m, r)$ with $(m, r) \neq (2, 2)$ and $n = (r^{2m} - 1)/(r - 1)$. (The case $(m, r) = (2, 2)$ leads to $S = A_6$ and $n = 15$, which can be excluded by Table 2.) Again, we have already considered the case $m = 1$, where $S \cong \mathrm{PSL}(2, r)$ (see [19, Theorem 8.1]). If $m = 2$ and $r = 3$, then $n = 40$ is not square free. The case $m \geq 3$ and $r = 2$ is listed as Case (c)(x) in our statement. In the remaining cases, n is the smallest index of any proper subgroup of S (see [13, Table 5.2.A]) and Remark 2.1 proves our claim.

If S is as in Line 12 or 13, then $S \cong A_6$, and we are done with Table 2.

Suppose that S is as in Line 14. Then $S = \Omega(2m + 1, r)$ and $n = (r^{2m} - 1)/(r - 1)$. We may assume that $m \geq 3$ and that r is odd, as otherwise $S \cong \mathrm{PSp}(2m, r)$ (see [19, Theorems 11.6, 11.9, Corollary 12.32]), a case already considered. If $r = 3$, then $n = (3^{2m} - 1)/2$ is not square free. In the other cases, n is the smallest index of any proper subgroup of S (see [13, Table 5.2.A]), and we are done as above.

Suppose that S is as in Line 16. Then, m is even and once more by [13, Table 5.2.A] and Remark 2.1 we obtain our claim. (This includes the case $m = 2$, where $S \cong \mathrm{PSL}(2, q^2)$ (see [19, Corollary 12.43]) and $n = q^2 + 1$.)

Suppose that S is as in Line 19. Then $S = \mathrm{P}\Omega^+(2m, r)$ with $m \geq 3$ odd and

$$n = \frac{(r^m - 1)(r^{m-1} + 1)}{r - 1}.$$

If $m = 3$, we have $S \cong \mathrm{PSL}(4, r)$ (see [19, Corollary 12.21]) and $n = (q^2 + 1)(q^2 + q + 1)$. This case is already contained in Case (c)(iii) of our statement. If $r \neq 2$ and $m \geq 5$, we conclude with [13, Table 5.2.A] and Remark 2.1. The case of $m \geq 5$ and $r = 2$ is included as Case (c)(xv) in our statement.

The remaining cases of [15, Table 3] are listed as Cases (c)(xi) through (c)(xiv) and (c)(xvi) through (c)(xvii), respectively, in our statement.

Suppose that S is as in [15, Table 4], i.e. an exceptional group of Lie type. In [21–23], A. V. Vasilyev lists the smallest index l of any proper subgroup of the exceptional simple groups. By Remark 2.1, the non-trivial conjugacy classes of G have at least l elements. We find $l = n$ except for

- $S = G_2(4)$ and $n = 1365$. We verified our claim for the two almost simple groups with socle $G_2(4)$ with GAP.

- $S = E_7(r)$ with $n = (r^{18} - 1)(r^{14} - 1)(r^4 - r^2 + 1)/((r^2 - 1)(r - 1))$. This case is listed as Case (c)(xviii) in our statement.

This completes our proof. □

Remark 2.3 For the purpose of this remark, let us call an *example* a pair (G, n) of a primitive permutation group G of square-free degree n containing a non-trivial conjugacy class with less than n elements.

Now, assume the hypotheses of Theorem 2.2. Clearly, not all the instances (G, n) listed there are examples.

If G is as in (a) of this theorem, then G has a conjugacy class of length $p - 1$. The symmetric group S_8 has a conjugacy class of length 28, and the sporadic simple group J_1 has a conjugacy class of length 1463. Thus the groups in (a) and the first three instances of (b) provide examples. This fact also indicates that in order to enumerate all examples one will have to use the classification of the finite simple groups.

The group $G = \mathrm{PGL}(2, r)$ has a conjugacy class of length $r(r - 1)/2$ for every odd prime r. The group $\mathrm{PGL}(2, 3)$ is isomorphic to the symmetric group S_4, which does not have a primitive permutation representation of degree 6. Thus $(\mathrm{PGL}(2, r), r(r + 1)/2)$ is an example, if and only if $r \geq 5$ and $(r + 1)/2$ is square free.

We expect that not many examples will arise from the pairs (G, n) listed in Theorem 2.2(c), but it would be a tedious task to enumerate all of them. One approach could be to determine all subgroups of G of index less than n and show that most of such subgroups are not centralizers of elements. Still, one has to decide whether one of the remaining numbers n is indeed square free. This will most certainly lead to difficult, if not intractible, number theoretical questions.

In the lemma below, we are going to make use of Zsigmondy primes, also known as primitive prime divisors. Let r and d be integers greater than 1. We call a prime ℓ a Zsigmondy prime for $r^d - 1$, if ℓ divides $r^d - 1$, but not $r^i - 1$ for $1 \leq i < d$. A Zsigmondy prime for $r^d - 1$ exists whenever $d > 2$ and $(r, d) \neq (2, 6)$ (see [10, Theorem IX.8.3]).

Lemma 2.4 *Suppose that $n = pq$, where p and q are distinct primes, and that G is a finite primitive permutation group of degree n such that G has a non-trivial conjugacy class with less than n elements. Then, one of the following holds:*

(a) *We have $G \in \{A_7, S_7, S_8\}$ and $n = 35$.*
(b) *We have $G = \mathrm{PGL}(2, r)$ for an odd prime r and $n = r(r + 1)/2$.*
(c) *The group G is almost simple with $\mathrm{PSL}(2, r) = \mathrm{soc}(G) \lneq G$, where $\sqrt{r}$ is an odd prime, $r > 9$, and $n = \sqrt{r}(r + 1)/2$.*
(d) *We have $G \in \{\mathrm{PSL}(2, 13), \mathrm{PGL}(2, 13)\}$ and $n = 91$.*
(e) *We have $G \in \{\mathrm{PSL}(2, 61), \mathrm{PGL}(2, 61)\}$ and $n = 1891$.*
(f) *We have $S = \mathrm{P}\Omega^+(2m, 2)$, for $m \geq 3$ and $n = (2^m - 1)(2^{m-1} + 1)$.*

Proof We have to exclude those integers n in Theorem 2.2 which are not the product of two different primes. From Cases (a) and (b) of Theorem 2.2, we obtain (part of) Case (a) and Case (b) of our lemma.

So, suppose that G is almost simple and that $S := \mathrm{soc}(G)$ occurs in Table 1. In Case (i), we have $n = \binom{c}{k}$ with $k \geq 3$. For reasons of symmetry, it suffices to consider $3 \leq k \leq c/2$. By [17, Theorem 7], the total number (counting multiplicities) of prime factors of the binomial coefficient $\binom{c}{k}$ is greater than or equal to the total number of prime factors of c, with equality only if $(c, k) = (8, 4)$. Thus, $\binom{c}{k}$ is the product of two different primes only if c is a prime. Consider the case $k = 3$ first. Then,

$$n = c \cdot \frac{(c-1)(c-2)}{6}.$$

If $c = 7$, we have $n = 35$. This is recorded in Case (a) of our lemma. So, assume that $c \geq 11$. Then, $n = \binom{c}{3}$ has at least three different prime factors. But then by [17, Theorem 3], the binomial coefficient $\binom{c}{k}$ with $k > 3$ also has at least three different prime factors.

In Cases (ii), (ix) through (xiv) and (xvi) through (xviii) of Table 1, the degree n is clearly not the product of two different primes.

In Case (iii) of Table 1, we have $\mathrm{soc}(G) = \mathrm{PSL}(m, r)$ and

$$n = \prod_{i=0}^{k-1}(r^{m-i} - 1)/\prod_{i=1}^{k}(r^i - 1),$$

with $2 \leq k < m$. For reasons of symmetry, it suffices to consider the integers k with $1 \leq k \leq m/2$. Suppose first that $m \geq 5$ and $k \geq 3$. Consider the terms $(r^m - 1)$, $(r^{m-1} - 1)$ and $(r^{m-2} - 1)$. They occur only in the numerator and not in the denominator of n and, by Zsigmondy's theorem, have pair-wise distinct primitive prime divisors r_1, r_2 and r_3 (which divide n) unless one of the pairs (m, r), $(m-1, r)$ or $(m-2, r)$ is equal to $(6, 2)$. But in these cases, i.e. if $m \in \{6, 7, 8\}$ and $r = 2$, we just compute that n is not the product of two different primes for all $3 \leq k \leq m/2$. If $m \geq 4$ and $k = 2$, then

$$n = \frac{(r^m - 1)(r^{m-1} - 1)}{(r^2 - 1)(r - 1)},$$

which is the product of two different primes if and only if $m \in \{4, 5\}$ and $r = 2$. But these cases have already been excluded.

In Case (iv) of Table 1, we have $S = \mathrm{PSL}(m, r)$ and

$$n = \prod_{i=0}^{2k-1}(r^{m-i} - 1)/(\prod_{i=1}^{k}(r^i - 1))^2,$$

with $m \geq 3$ and $1 \leq k < m/2$. Suppose first we have $m \geq 5$ and $k \geq 2$. Then, again, the terms $(r^m - 1)$, $(r^{m-1} - 1)$ and $(r^{m-2} - 1)$ occur only in the numerator and not in the denominator of n and have pair-wise distinct primitive prime divisors r_1, r_2 and r_3 (which divide n) unless one of the pairs (m, r), $(m-1, r)$ or $(m-2, r)$ is equal to (6, 2). But in these cases, i.e. if $m \in \{6, 7, 8\}$ and $r = 2$, we just compute that n is not the product of two different primes for all $2 \leq k < m/2$. If m is arbitrary and $k = 1$ then,

$$n = \frac{(r^m - 1)(r^{m-1} - 1)}{(r - 1)(r - 1)}$$

which is the product of two different primes if and only if $(m, r) = (3, 2)$. But this case has already been excluded.

Case (v) of Table 1 is listed as Case (c) in our lemma.

In Cases (vi) through (viii) of Table 1, we have $n = r(r^2 - 1)/d$ with $d \in \{24, 48, 120\}$. Clearly, n is not the product of two different primes if $r > d + 1$. For $r \leq d + 1$ and r not equal to one of the primes excluded in Table 1, we have that n is the product of two different primes if and only if $(r, d) = (13, 24)$ or $(r, d) = (61, 120)$. We have $S = \mathrm{PSL}(2, r)$, and as r is a prime, $G = \mathrm{PSL}(2, r)$ or $G = \mathrm{PGL}(2, r)$. These cases are listed as Case (d) and Case (e) of our lemma.

Finally, Case (xv) of Table 1 is Case (f) of the lemma. □

3 The Action of the Right Multiplication Groups of Rcc Loops of Twice Prime Order

The results of the previous section are now applied to RCC loops whose order is the product of two distinct primes. Recall the notion of the envelope of a loop as introduced in the second paragraph of Sect. 1 (which follows [2, p. 100]). Recall that a loop folder is a triple (G, H, T), where G is a finite group, H is a subgroup of G and T is a transversal, with $1 \in T$, for all coset spaces $H^g \backslash G$ with $g \in G$ (see the third paragraph of Sect. 1, which follows [2, p. 101]). A loop folder (G, H, T) is an RCC loop folder, if T is invariant under conjugation by G (see also Sect. 1). We will now introduce further notation, although only needed in later sections.

Let (G, H, T) be a loop folder. By definition, the *order* of (G, H, T) is the size of T. We say that (G, H, T) is *faithful*, if G acts faithfully on $H \backslash G$. This is the case if and only if the core of H in G is trivial. Recall that the smallest normal subgroup C of G contained in H is called the *core of* H *in* G. Thus, C is the intersection of all the G-conjugates of H in G, i.e. $C := \bigcap_{g \in G} H^g$. The core of H in G is equal to the kernel of the permutation representation of G on the (right or left) cosets of H. Clearly, the envelope of a loop is a faithful loop folder.

Here is the main result of this section. It is only used in the setup of Sect. 5.1, and nowhere else in this paper.

Theorem 3.1 *Let (G, H, T) denote the envelope of an RCC loop of order $n = pq$, where p and q are distinct primes. Then, G acts imprimitively on $H \backslash G$.*

Proof Let $n = pq = |H \backslash G|$. Suppose that G acts primitively on $H \backslash G$. Since $|T| = n$ and T is a union of conjugacy classes one of which is the trivial class, G has a non-trivial conjugacy class with less than n elements. Hence, G is one of the groups of Lemma 2.4.

In Cases (a), (b), (d) and (e) of Lemma 2.4, the concerned groups have at most two non-trivial conjugacy classes with less than n elements. Elementary combinatorics shows that in these cases, there is no union of conjugacy classes T with $|T| = n$.

Suppose that G is as in Case (c) of Lemma 2.4. Then, G is almost simple with $S := \mathrm{soc}(G) = \mathrm{PSL}(2, r)$, where $\sqrt{r}$ an odd prime, $r > 9$, and $n = \sqrt{r}(r + 1)/2$. Moreover, $S \lneq G$. The subgroups of $S = \mathrm{PSL}(2, r)$, are classified in Dickson's Theorem; see [11, Hauptsatz II.8.27]. This shows that if $r \geq 17$, then only the maximal subgroups of index $r + 1$ have an index less then $n = \sqrt{r}(r + 1)/2$. Consider the non-trivial conjugacy classes of G. Those which contain elements of S have at least n elements as we already mentioned in the proof of Theorem 2.2.

Let $g \in G \setminus S$. By Remark 2.1 we have

$$|G\colon C_G(g)| = a \cdot |S\colon C_S(g)|$$

for some positive integer a. Hence, $|G\colon C_G(g)| \geq n$ except if $C_S(g) \leq M$, where M is a maximal subgroup of S with $|S\colon M| = r + 1$. In this case, $r + 1 \mid |S\colon C_S(g)|$. Hence, if $|G\colon C_G(g)|$ is less than n, it is a multiple of $r + 1$. Thus, a union T of conjugacy classes of sizes less than n with $1 \in T$ has a size congruent to 1 modulo $r + 1$ and is therefore not divisible by the prime $(r + 1)/2$. Therefore, it is not possible to have $|T| = n$.

Finally, assume that G is as in Case (f) of Lemma 2.4. Then G is almost simple with $S := \mathrm{soc}(G) = \mathrm{P}\Omega^+(2m, 2)$, where $m \geq 3$ and $n = (2^m - 1)(2^{m-1} + 1)$. In order for n to be the product of two distinct primes, it is necessary that $p := 2^{m-1} + 1$ is a Fermat prime and $q := 2^m - 1$ is a Mersenne prime. In particular, m is a Fermat prime. By [13, Table 5.2.A], the smallest index of any proper subgroup of S equals $2^{m-1}(2^m - 1)$. Remark 2.1 shows that any non-trivial conjugacy class of G has at least $2^{m-1}(2^m - 1)$ elements. As twice this number is greater than n, we conclude that T is the union of two conjugacy classes of G, one of which has length $2(2^{2m-2} + 2^{m-2} - 1)$. We have $m \geq 5$ and $S = \mathrm{SO}^+(2m, 2)$ (in the notation of [19, p. 160]). In particular, $G \in \{\mathrm{SO}^+(2m, 2), O^+(2m, 2)\}$. For the order of G, see [19, p. 141, 165].

Let ℓ be a Zsigmondy prime for $2^{2m-2} - 1$. As ℓ does not divide $2^{m-1} - 1$, and as $p = 2^{m-1} + 1$, we conlcude that $\ell = p$. Let t be a non-trivial element in T. Now, p does not divide $2(2^{2m-2} + 2^{m-2} - 1) = |G\colon C_G(t)|$, and so there is $g \in C_G(t)$ with $|g| = p$. Let V denote the natural $2m$-dimensional $\mathbb{F}_2$-vector space of G, equipped with the quadratic form Q defining G. Since p is a Zsigmondy prime for $2^{2m-2} - 1$, it follows that g acts irreducibly on some $(2m - 2)$-dimensional subspace V_0 of V. As the dimension of V_0 is larger than 1, either V_0 is totally singular or non-degenerate with respect to Q (for these notions see [19, p. 56]). The maximal dimension of a totally singular subspace of V equals m, and thus V_0 is in fact non-degenerate. It follows that g fixes $V_1 := V_0^{\perp}$. In particular, $g \in O(V_0) \times O(V_1)$, where $O(V_i)$ denotes the orthogonal group with respect to the restriction Q_i of Q to V_i, $i = 0, 1$. We may thus write $g = g_0 \oplus g_1$, whith g_i the restriction of g to V_i, $i = 0, 1$.

Now $O(V_0)$ contains a cyclic, irreducible subgroup, and thus the Witt index of Q_0 equals $m-2$ by [12, Satz 3c)]. Hence V_1 does not contain any non-trivial singular vector with respect to Q_1, and thus $O(V_1) \cong S_3$, the symmetric group on three letters (see [19, Theorem 11.4]). As $|g| = p \geq 17$, we conclude that g acts trivially on V_1, i.e. $g_1 = 1$ and V_1 is the fixed space of g. It follows that $C_G(g)$ fixes V_1 and $V_0 = V_1^{\perp}$, and thus $C_G(g) \leq C_{O(V_0)}(g_0) \times O(V_1)$. As g_0 acts irreducibly on V_0, its centralizer in $O(V_0)$ is cyclic and irreducible, and thus equals $\langle g_0 \rangle$, again by [12, Satz 3c)]. In particular, $t \in C_G(g) \leq \langle g_0 \rangle \times O(V_1)$. If $p \mid |t|$ or $3 \mid |t|$, then $C_G(t) \leq O(V_0) \times O(V_1)$. Otherwise, $|t| = 2$ and t has a $(2m-1)$-dimensional fixed space V' on V and $C_G(t) \leq O(V')$. In any case, the 2-part of $|C_G(t)|$ is less than $2^{(m-1)^2}$, whereas the 2-part of $|G|$ equals $2^{m(m-1)+1}$ (see [19, p. 141]). In particular, the 2-part of $|G\colon C_G(t)|$ is larger than 2, a contradiction. □

We end this section with two general results on loop folders with certain invariance properties. The first will be used in an extension of a theorem of Drápal [9].

Lemma 3.2 *Let (G, H, T) be a loop folder such that T is invariant under conjugation by H. Suppose the $t \in T$ is such that Ht is also a left H-coset in G, i.e. there exist $g \in G$ with $Ht = gH$ (this is the case in particular if t normalizes H). Then $[t, H] = 1$.*

Proof Let $h \in H$. Then

$$Ht = gH = gHh = Hth = Hh^{-1}th.$$

This implies $t = h^{-1}th$, as $t, h^{-1}th \in T$. □

Lemma 3.3 *Let (G, H, T) denote an RCC loop folder. Let $K \leq G$ such that $HG' \leq K$. Then,*

$$|G\colon C_G(t)| \leq |K\colon H| \quad \textit{for all } t \in T$$

(i.e. the length of the conjugacy classes of the elements in T are bounded above by $|K\colon H|$).

Proof Let $g \in G$. Then, the right coset Kg is a union of exactly $|K\colon H|$ right cosets of H in G. Thus, $|Kg \cap T| = |K\colon H|$.

Now, let $t \in T$ and $x \in G$. Then, $t^x t^{-1} \in G' \leq K$. It follows that $Kt^x = Kt$ and thus $t^x \in Kt$ for all $x \in G$. As $t^x \in T$ for all $x \in G$ by assumption, we conclude that $|G\colon C_G(t)| = |\{t^x \mid x \in G\}| \leq |K\colon H|$. □

4 Right Conjugacy Closed Loops of Prime Order

In this section, we give a new proof of a theorem of Drápal [9] which states that left conjugacy closed loops of prime order are groups. We prove the analogue for RCC loops, but as the opposite loop of a left conjugacy closed loop is an RCC loop, our version is equivalent to Drápal's result. Recall the notions related to loop folders summarized at the beginning of Sect. 3.

We begin with an easy lemma.

Lemma 4.1 *Let p be a prime and let $G \leq S_p$ with $p \mid |G|$. Then, the following statements hold for every $1 \neq g \in G$.*

(a) *If $p \nmid |g|$, then $p \mid |G\colon C_G(g)|$.*
(b) *If $p \mid |g|$, then $|C_G(g)| = p$.*
(c) *Suppose that $|G\colon C_G(g)| < p$. Then G has a unique Sylow p-subgroup P and $g \in P$. Moreover, if $G \neq P$, then G is a Frobenius group with kernel P and a Frobenius complement of order r dividing $p-1$. In this case, P is the Frattini subgroup of G.*
(d) *We have $|G\colon C_G(g)| \neq 2(p-1)$.*

Proof In view of the cycle decomposition of g, the first two parts are trivial. So let us assume that $|G\colon C_G(g)| < p$ or that $|G\colon C_G(g)| = 2(p-1)$. By (a) and (b), we have $|C_G(g)| = p$, and in particular $P := \langle g \rangle$ is a Sylow p-subgroup of G. Under the hypothesis of (c), we get $|G\colon N_G(P)| < p$, and under the hypothesis of (d) we get $|G\colon N_G(P)| \mid 2(p-1)$. In each case, Sylow's theorems imply $P \trianglelefteq G$. Now (d) and the last two statements of (c) follow from the fact that G embeds into $N_{S_p}(P)$, which is isomorphic to $\mathrm{Aff}(1, p)$. □

Corollary 4.2 (Drápal [9]) *Let p be a prime and let (G, H, T) denote the envelope of an RCC loop $\mathcal{L}$ of order p. Then, $H = 1$, i.e. $\mathcal{L}$ is a group (isomorphic to G).*

Proof We may assume that $G \leq S_p$ and we have $p \mid |G|$. Now $T = \{1\} \cup T'$ with $T' := T \setminus \{1\}$. By assumption, T' is a union of conjugacy classes of G of lengths at most $p-1$. It follows from Lemma 4.1(c) that G has a unique Sylow p-subgroup P and that $T \subseteq P$. Hence, $G = \langle T \rangle = P$, i.e. $H = 1$. □

We will also need the following generalization of Corollary 4.2.

Proposition 4.3 *Let p be a prime and let (G, H, T) be an RCC loop folder of order p with $\langle T \rangle = G$. Then, G is abelian.*

Proof Let

$$N := \bigcap_{g \in G} H^g$$

denote the kernel of the action of G on $H \backslash G$ and put $\bar{G} := G/N$, $\bar{H} := H/N$ and $\bar{T} := \{Nt \mid t \in T\} \subseteq \bar{G}$. Then $(\bar{G}, \bar{H}, \bar{T})$ is a faithful RCC loop folder of order p with $\langle \bar{T} \rangle = \bar{G}$. Thus, $(\bar{G}, \bar{H}, \bar{T})$ is the envelope of an RCC loop of order p (see [2, 1.7.(4)]). By the result of Drápal (see Corollary 4.2), such a loop is a group. It follows that $\bar{T} = \bar{G}$ and $\bar{H} = 1$. In particular, $H = N$ is a normal subgroup of G of index p.

To show that G is abelian, let $t \in T$. By Lemma 3.2 we have $[t, H] = 1$, as $H \trianglelefteq G$. It follows that $\langle H, t\rangle \leq C_G(t)$. If $t \neq 1$, we have $\langle H, t\rangle = G$, as H is of index p in G. Hence $t \in Z(G)$ for all $t \in T$. The claim follows from $\langle T\rangle = G$. □

Notice that the above proposition is a generalization of Drápal's theorem (see Corollary 4.2); indeed, if (G, H, T) is the envelope of a loop, then $G = \langle T\rangle$, and if, moreover, G is abelian, then $H = 1$, as the core of H in G is trivial.

5 The Right Multiplication Groups of Rcc Loops of Twice Prime Order

We refer the reader to the introduction of Sect. 3 for the notions related to loop folders.

5.1 *Generalities*

Let p and q be distinct primes and let (G, H, T) denote the envelope of an RCC loop of order pq. This implies in particular that G acts faithfully on $H\backslash G$, i.e. the core of H in G is trivial. It is at this stage, and only here, where we impose an important consequence of Theorem 3.1. This states that G acts imprimitively on $H\backslash G$, and hence H is not a maximal subgroup of G. We let $K \lneq G$ such that $H \lneq K$. We choose notation such that $|G\colon K| = q$ and $|K\colon H| = p$.

Put $T_1 := T \cap K$, $K_1 := \langle T_1\rangle \leq K$ and $H_1 := H \cap K_1 \leq H$.

We collect first properties.

Lemma 5.1 *Let the notation be as above. Then, (K_1, H_1, T_1) is an RCC loop folder of order p with K_1 abelian. Also, $K_1 \trianglelefteq K$ and $K = HK_1$. Finally, $H_1 \trianglelefteq K$.*

Proof Clearly, K is the disjoint union of the cosets Ht for $t \in T_1$. Thus $|T_1| = p$ and K_1 is the disjoint union of the cosets H_1t for $t \in T_1$. As T_1 is invariant under conjugation in K, the first statement follows. The second statement follows from Proposition 4.3, and the next two statements are obvious. The last statement follows from $K = HK_1$ and the fact that $H_1 \trianglelefteq H$ and that K_1 is abelian. □

5.2 *The Case $q = 2$*

Let us assume throughout this subsection that $q = 2$. Then $K \trianglelefteq G$. Moreover, $K_1 \trianglelefteq G$, as $K_1^g = \langle (T \cap K)^g\rangle = \langle T \cap K\rangle = K_1$ for all $g \in G$.

Lemma 5.2 *Let $L \leq H$ with $L \trianglelefteq K$. Then $L \cap L^a = 1$ and $LL^a = L \times L^a \trianglelefteq G$ for all $a \in G \setminus K$.*

Proof Let $a \in G \setminus K$. Clearly, $L \cap L^a$ and LL^a are normal subgroups of G, as $a^2 \in K$, and $L \trianglelefteq K$. Thus $L \cap L^a = 1$ since $L \cap L^a \leq H$ and the core of H in G is trivial. As $L^a \trianglelefteq K$, the product LL^a is direct. □

We record two consequences which will be used later on.

Corollary 5.3 *We have $H_1 \in \{1, p\}$ and K_1 is elementary abelian of order p or p^2.*

Proof If H_1 is trivial, K_1 has order p by Lemma 5.1. Suppose that H_1 is non-trivial and let $a \in T \setminus K$. Then, $H_1 \cap H_1^a = 1$ by Lemma 5.2. As $K_1 \trianglelefteq G$, we have $H_1 H_1^a \leq K_1$. It follows that $|H_1|^2$ divides $|K_1| = p|H_1|$. This implies $|H_1| = p$ and $K_1 = H_1 H_1^a$, yielding our claim. □

Corollary 5.4 *Suppose that $1 \neq H \trianglelefteq K$. Then, $|H| = p$ and there is an involution $a \in G \setminus K$ such that $K = H \times H^a$. In particular, G is isomorphic to the wreath product $C_p \wr C_2$.*

Proof By Lemma 5.2 we have $H \cap H^a = 1$ and $HH^a \leq K$ for every $a \in G \setminus K$. As $|K| = p|H|$, this implies that $|H| = p$ and $K = H \times H^a$. It also follows that the involutions in G are contained in $G \setminus K$ and thus $G \cong C_p \wr C_2$. □

We now distinguish two cases.

5.2.1 Case 1

Assume that $H \neq 1$ and that $[s, K_1] = 1$ for all $s \in T \setminus K$.

Proposition 5.5 *Under the assumptions of Sect. 5.2.1, we have $H_1 = 1$ and $K_1 \leq Z(G)$. Moreover, G is isomorphic to the wreath product $C_p \wr C_2$.*

Proof The fact that K_1 is abelian and our hypothesis imply that $T \subseteq C_G(K_1)$, and hence $G = \langle T \rangle \leq C_G(K_1)$, i.e. $K_1 \leq Z(G)$. Thus, $H_1 \trianglelefteq G$, which implies $H_1 = 1$, as $H_1 \leq H$ and the core of H in G is trivial. Now $K = HK_1$ by Lemma 5.1, and thus $H \trianglelefteq K$. The claim follows from Corollary 5.4. □

5.2.2 Case 2

Assume that there is $s \in T \setminus K$ with $[s, K_1] \neq 1$. In this case, we put $Z := K_1 \cap C_G(s)$. Also, we let $C := \bigcap_{k \in K} H^k \trianglelefteq K$ denote the kernel of the action of K on the cosets of H in K.

Lemma 5.6 *Under the assumptions and with the notation of Sect. 5.2.2, the following statements hold.*

(a) *We have $|G : C_G(s)| = p$ and $|Z| = |H_1|$.*
(b) *We have $G = C_G(s)K_1$ and $Z \leq Z(G)$.*

(c) *The centralizer $C_G(s)$ is abelian and $C_G(s)/Z$ is cyclic.*

Proof By assumption, $K_1 \not\leq C_G(s)$. Corollary 5.3 implies that $|Z| \in \{1, p\}$. By Lemma 5.1, we have $|T_1| = p$, which implies that $|T \setminus K| = p$ (recall that $q = 2$ and thus $|T| = 2p$). As $T \setminus K$ is a union of conjugacy classes of G, we have $|G : C_G(s)| \leq p$, i.e. $|C_G(s)| \geq |G|/p$.

Now if $H_1 = 1$, i.e. $|K_1| = p$, we also have $|Z| = 1$ and $G = C_G(s)K_1$. Thus all statements of (a) and (b) hold in this case.

Now, assume that $|H_1| = p$, i.e. K_1 is elementary abelian of order p^2. Then, $|Z| \leq p$ and

$$|G| \geq |C_G(s)K_1| \geq \frac{|C_G(s)||K_1|}{|Z|} \geq \frac{|G|}{p} \cdot \frac{p^2}{|Z|} \geq |G|,$$

and we must have equality everywhere in the above chain of inequalities. This implies $|Z| = p$ and $|C_G(s)K_1| = |G|$, again yielding all the claims of (a) and the first claim of (b).

In any case, the set $T \setminus K$ is a conjugacy class of G, consisting of the elements $\{s^k \mid k \in K_1\}$. Write $\bar{\ }: G \to \bar{G} := G/K_1$ for the canonical epimorphism. We have $\bar{G} = \langle \bar{T} \rangle = \langle \bar{s} \rangle$ as $T_1 \subseteq K_1$. The natural isomorphism $\bar{G} \to C_G(s)/Z$ maps $\bar{s}$ to $Zs \in C_G(s)/Z$. Thus $C_G(s)/Z = \langle Zs \rangle$ and $C_G(s) = \langle Z, s \rangle$. In particular, $C_G(s)/Z$ is cyclic and $C_G(s)$ is abelian, as $Z \leq C_G(s)$. Now $G = C_G(s)K_1$ and $Z = K_1 \cap C_G(s)$ imply that $Z \leq Z(G)$. □

The previous result implies, in particular that G is soluble. Indeed, K_1 is a normal subgroup of G, as we have remarked at the beginning of Sect. 5.2.2. Now, K_1 is abelian by Corollary 5.3, and $G/K_1 = C_G(s)K_1/K_1 \cong C_G(s)/Z$ by Part (b) of the lemma above and by the definition of Z. By Part (c) of the lemma, $C_G(s)/Z$ is cyclic, hence G is soluble.

Corollary 5.7 *Let the assumptions and notation be as in Sect. 5.2.2. If $H_1 \neq 1$, then $H \trianglelefteq K$.*

Proof Suppose that $H_1 \neq 1$. Then $|Z| = p$ by Corollary 5.3 and Lemma 5.6(a). Now $Z \leq Z(G)$ by Lemma 5.6(b), and thus $Z \cap H = 1$, since the core of H in G is trivial. It follows that $K = H \times Z$, as $|K : H| = p$ and $Z \leq K_1 \leq K$.

Now, K/C is a soluble permutation group on p points. It follows from a theorem of Galois (see [11, Satz II.3.6]) that K/C is isomorphic to a subgroup of the affine group $\mathrm{Aff}(1, p)$.

We have $K/C = (H \times Z)/C = H/C \times ZC/C \cong H/C \times Z$. This implies that H/C is trivial, i.e. $H = C \trianglelefteq K$. □

Lemma 5.8 *Let the assumptions and notation be as in Sect. 5.2.2. If $H_1 = 1$, then $C = 1$ or H is a p-group.*

Proof Put $L := C_G(s)$. Then L is cyclic, $G = L \ltimes K_1$ and $K = H \ltimes K_1$, with $|K_1| = p$ (see Corollary 5.3 and Lemma 5.6). In particular, H is abelian, as it is isomorphic to a subgroup of L.

Let ℓ be a prime different from p and let $S \leq H$ denote a Sylow ℓ-subgroup of G. As $|G: L| = p$, there is $g \in G$ such that $S^g \leq L$. As L is abelian, we have $L \leq C_G(S^g)$. Suppose that $C_G(S^g) = G$, i.e. $S^g \leq Z(G)$. Then $S \leq Z(G)$ wich implies $S = 1$, as the core of H in G is trivial.

Now, assume that $\ell \mid |H|$. By the above, we must have $C_G(S^g) = L$. Then, $C_G(S)$ is a cyclic complement of K_1 in G containing H. As $C \leq H \leq C_G(S)$ and $C \trianglelefteq K$, it follows that $C \trianglelefteq G$ und thus $C = 1$. □

Corollary 5.9 *Let the assumptions and notation be as in Sect. 5.2.2. If $H_1 = 1$, then $C = 1$ or $H \trianglelefteq K$.*

Proof Suppose that $C \neq 1$. Then $K = HK_1$ is a p-group by Lemma 5.8. As K/C is isomorphic to a subgroup of the affine group $\mathrm{Aff}(1, p)$ by Galois' theorem (see [11, Satz II.3.6]), it follows that $|K/C| = p$. Now $C \cap C^s = 1$ and $CC^s = C \times C^s \trianglelefteq G$ by Lemma 5.2. Thus $C^s \cong (C \times C^s)/C \leq K/C$, and hence C has order p. Therefore, $|K| = p^2$ and hence K is abelian, proving our claim. □

5.3 *The Case $p = 2$*

Let us assume throughout this subsection that $p = 2$. Then $H \trianglelefteq K$. Here, we put $D := \bigcap_{g \in G} K^g$, the kernel of the action of G on the set of right cosets of K. Then, D is an elementary abelian 2-group, as G acts faithfully on the set of right cosets of H and as H has index 2 in K. We write $^- : G \to \bar{G} := G/D$ for the canonical epimorphism. Then, $\bar{G}$ is a faithful permutation group on q letters, i.e. $\bar{G}$ is isomorphic to a subgroup of S_q.

Lemma 5.10 *We have $|K \cap T| = 2$, and writing $K \cap T = \{1, z\}$, we have $z \in Z(K)$. In particular, $C_G(z) \in \{K, G\}$.*

Proof The first assertion follows from $|K: H| = 2$, and the second from the fact that all K-conjugates of z again are in $K \cap T$. □

5.3.1 Case 1

Here, we consider the case that K is normal in G. Let us keep the notation of Lemma 5.10 in the following.

Lemma 5.11 *Suppose that $K \trianglelefteq G$. Then, G is abelian and $H = 1$.*

Proof In this case $z^g \in K \cap T$ for all $g \in G$, and thus $\langle z \rangle \leq Z(G)$. Now, $\langle z \rangle \cap H = 1$, as the core of H in G is trivial. It follows that $|z| = 2$ and $K = H \times \langle z \rangle$. This implies that $H' = K' \trianglelefteq G$, and thus $K' = 1$, i.e. K is abelian. In turn, K is a 2-group, as $O_{2'}(H) = O_{2'}(K) \trianglelefteq G$ (recall that $O_{2'}(H)$ denotes the largest normal subgroup of H of odd order). Thus, K is the unique Sylow 2-subgroup of G.

Now, let $t \in T \setminus K$. Then, t is not a 2-element as otherwise $t \in K$. Let r be an integer such that $Q := \langle t^r \rangle$ is a Sylow q-subgroup of G. As G/K is cyclic, we have $G' \leq K$ and we may thus apply Lemma 3.3. This yields $|G\colon C_G(t)| \leq 2$. Hence $|G\colon C_G(Q)| \leq 2$ and thus $C_G(Q) \trianglelefteq G$. Now, Q is abelian and hence $Q \leq C_G(Q)$. It follows that $Q \trianglelefteq C_G(Q)$ and thus $Q = O_q(C_G(Q))$. In particular, $Q \trianglelefteq G$, implying that $G = Q \times K$ is abelian. □

5.3.2 Case 2

Here, we consider the case that K is not normal in G. Again, we use the notation of Lemma 5.10.

Proposition 5.12 *Suppose that $N_G(K) = K$. Then there is $M \trianglelefteq G$ such that $|G\colon HM| = 2$.*

Proof Let Z denote the G-conjugacy class of z and put $T' := T \setminus (Z \cup \{1\})$. By Lemma 5.10, we have $C_G(z) \in \{K, G\}$. Suppose first that $C_G(z) = K$. Then $|Z| = |G\colon K| = q$, and thus $|T'| = q - 1$. If $C_G(z) = G$, then $|T'| = 2(q-1)$. In particular, $q - 1 \mid |T'|$. Let $X_1, \ldots, X_m$ denote the G-conjugacy classes contained in T', numbered in such a way that $|X_1| \leq \cdots \leq |X_m|$. Thus $|X_1| \leq q - 1$, unless $m = 1$ and $C_G(z) = G$, in which case $|X_1| = 2(q-1)$.

Let $t \in T'$. Then $t \notin K$, as $t \notin \{1, z\} = K \cap T$. In particular, $\bar{t} \neq 1$, since $D \leq K$. Let X denote the G-conjugacy class of t. Then, $\bar{X}$ is the $\bar{G}$-conjugacy class of $\bar{t}$ and $|\bar{X}|$ divides $|X|$. Consider the case that $X = X_1$. If $|X| \leq q - 1$, then $|\bar{X}| \leq q - 1$. If $|X| = 2(q-1)$, then $|\bar{X}|$ is a proper divisor of $2(q-1)$ by Lemma 4.1(d), and thus, again, $|\bar{X}| \leq q - 1$. Lemma 4.1(c) implies that $\bar{G}$ has a normal Sylow q-subgroup Q. Moreover, $|\bar{G}\colon \bar{K}| = |G\colon K| = q$, and $\bar{K} \neq 1$, as otherwise $K = D$ would be normal in G. Thus, $\bar{G}$ is a Frobenius group of order qr with $r \mid (q-1)$, again by Lemma 4.1(c).

Since $\bar{G}$ is a Frobenius group, every non-trivial conjugacy class of $\bar{G}$ has length q or r, and the conjugacy classes of length r lie in Q. Suppose that there is some $1 \leq j \leq m$ such that $|\bar{X}_j| = q$. Then, $|X_j| = q$ as $|\bar{X}_j|$ divides $|X_j|$ and $|X_j| \leq 2(q-1)$. Also, X_j is the unique conjugacy class of length q contained in T'. If $1 \leq i \neq j \leq m$, then $|X_i| \leq q - 1$, and thus $|\bar{X}_i| = r$. In particular, $r \mid |X_i|$. Now, $q - 1$ divides $|T'|$, as we have already observed above. It follows that r divides q, a contradiction. This shows that $\bar{T}' \subseteq Q$.

We have $z \in Z(K)$ and $D \leq K$, and thus $Z \subseteq C_G(D)$ as $D \trianglelefteq G$. Moreover, D is abelian and hence $\langle Z, D \rangle \leq C_G(D)$. Now, $\bar{G} = \langle \bar{T} \rangle = \langle \bar{Z} \cup \bar{T}' \rangle = \langle \bar{Z} \rangle$, as $\bar{T}' \subseteq Q$, the Frattini subgroup of $\bar{G}$. Thus, $G = \langle Z, D \rangle \leq Z_G(D)$, i.e. $D \leq Z(G)$. This implies that $D \cap H = 1$, as the core of H in G is trivial. Hence, $|H||D| = |HD| \leq |K| = 2|H|$, and so $|D| \in \{1, 2\}$.

Let M denote the inverse image of Q in G. If $D = 1$, then $|M| = q$ an thus $G = K \ltimes M$ and $|G\colon HM| = 2$ as claimed. Now suppose that $D = \langle d \rangle$ with $|d| = 2$. Then $HM \neq G$ as otherwise $d \in HM$ and, in turn, $d \in H$. Now, $G = KM$ as $\bar{G} = \bar{K}\bar{M}$, and thus $G = KM = HM \cup HMd$, i.e. $|G\colon HM| = 2$, as claimed. □

5.4 The Main Result

We can now summarize our results for envelopes of RCC loop folders of orders $2p$ for odd primes p.

Theorem 5.13 *Let (G, H, T) be the envelope of an RCC loop of order $2p$, where p is an odd prime. Then, there is a subgroup $K \leq G$ with $H \leq K$ and $|G : K| = 2$ and one of the following occurs.*

(a) *The group G is isomorphic to the wreath product $C_p \wr C_2$.*
(b) *The group G is isomorphic to a subgroup of the affine group* $\mathrm{Aff}(1, p)$.
(c) *We have $G = K \times \langle a \rangle$, and K has odd order and is isomorphic to a subgroup of the affine group* $\mathrm{Aff}(1, p)$.

In Cases (b) *and* (c), *$\langle T \cap K \rangle$ is a normal subgroup of G of order p. The Cases* (a), (b) *and* (c) *are disjoint.*

Proof The first statement follows from Lemma 5.11 and Proposition 5.12 (with q replaced by p). In particular, we are in the situation of Sect. 5.2.

In the following, we resume to the notation introduced at the beginning of Sect. 5.1. Suppose that G is not isomorphic to the wreath product $C_p \wr C_2$. By Proposition 5.5, we may assume that we are in the situation of Sect. 5.2.2. Corollary 5.4 implies that H is not a normal subgroup of K. Hence $H_1 = 1$ and $C = 1$ by Corollaries 5.7 and 5.9. In particular, $|K_1| = p$ by Corollary 5.3.

If $C_G(K_1) = K_1$, then G/K_1 injects into the automorphism group of K_1, and thus G is as in (b). Assume now that $K_1 \lneq C_G(K_1)$. As $K = HK_1$ by Lemma 5.1, we have $C_H(K_1) \leq C$, and thus $H \cap C_G(K_1) = C_H(K_1) = 1$. Hence $|HC_G(K_1)| = |H||C_G(K_1)| > |H||K_1| = |K|$, and thus $HC_G(K_1) = G$ and $|C_G(K_1)| = 2|K_1|$. It follows that $C_G(K_1) = K_1 \times \langle a \rangle$ for some $a \in G \setminus K$ of order 2. As $K \trianglelefteq G$, and $\langle a \rangle = O_2(C_G(K_1)) \trianglelefteq G$, we have $G = K \times \langle a \rangle$. Now K acts faithfully on the set of H-cosets in K, and thus $K = HK_1$ is isomorphic to a subgroup of the affine group $\mathrm{Aff}(1, p)$. Finally, K has odd order since G/K_1 is cyclic by Lemma 5.6. □

6 The Rcc Loops of Twice Prime Order

Let p be an odd prime. In this section, we determine the number of isomorphism classes of RCC loops of order $2p$. Let $\mathcal{L}$ denote such a loop and let (G, H, T) be its envelope. By numbering the elements of $\mathcal{L}$ by the integers $1, \ldots, 2p$, where 1 numbers the identity element of $\mathcal{L}$, we may and will view G as a subgroup of S_{2p}, and H as the stabilizer in G of 1. If $\mathcal{L}_1$ and (G_1, H_1, T_1) is another such configuration, then $\mathcal{L}$ and $\mathcal{L}_1$ are isomorphic as loops, if and only if there is an element of S_{2p}, conjugating (G, H, T) to (G_1, H_1, T_1). The isomorphism types of the right multiplication groups arising in RCC loops of order $2p$ have been described in Theorem 5.13. For each of these groups G we have to determine their embeddings into

S_{2p} up to conjugation. This will yield the possible pairs (G, H) to be considered. For each of these pairs, we have to determine the normalizer N in S_{2p} of G and H, and then find the distinct N-orbits of G-invariant transversals T for $H\backslash G$ such that $1 \in T$ and $\langle T \rangle = G$. We will refer to the three different types of G in Theorem 5.13(a), (b) and (c) as Case (a), (b) and (c), respectively.

We begin with some preliminary results. As usual, the largest normal p-subgroup of a finite group U is denoted by $O_p(U)$, and its largest normal subgroup of odd order by $O_{2'}(U)$.

Lemma 6.1 *Let $\pi_1, \alpha \in S_{2p}$ be defined by $\pi_1 := (1, 2, \ldots, p)$ and $\alpha := (1, p+1)(2, p+2)\cdots(p, 2p)$. Put $\pi_2 := \pi_1^\alpha = (p+1, p+2, \ldots, 2p)$. Let $\nu_1 \in S_p$ be an element of order $p-1$ such that $N_{S_p}(\langle \pi_1 \rangle) = \langle \pi_1, \nu_1 \rangle$. Put $\nu_2 := \nu_1^\alpha$ and $\nu := \nu_1\nu_2$.*

(a) *Let $G := \langle \pi_2, \alpha \rangle$. Then $Z(G) = \langle \pi_1\pi_2 \rangle$, $G' = \langle \pi_1^{-1}\pi_2 \rangle$ and $G = C_{S_{2p}}(Z(G)) \cong C_p \wr C_2$. Put $N := N_{S_{2p}}(G)$. Then $N = N_{S_{2p}}(Z(G)) = \langle \nu \rangle \ltimes G$.*
(b) *Let $U \leq N$ with $O_p(U) = Z(G)$ and $U \nleq G$. Then, there is $n \in N$ such that $N_{S_{2p}}(U^n) = A \times \langle \alpha \rangle$ with $A = \langle \nu \rangle \ltimes Z(G)$.*

Proof (a) The statements about G are trivially verified. From $G = C_{S_{2p}}(Z(G))$ we conclude that $G \trianglelefteq N_{S_{2p}}(Z(G))$, and thus $N = N_{S_{2p}}(Z(G))$. Moreover, N/G is isomorphic to a subgroup of $\mathrm{Aut}(Z(G))$, which is a cyclic group of order $p-1$. Now $\langle \nu \rangle \cap G = 1$, as the elements in G fixing the set $\{1, \ldots, p\}$ have order divisible by p. Also, ν normalizes G, and thus $N = \langle \nu \rangle \ltimes G$.

(b) From $O_p(U) = Z(G)$ we conclude that $N_{S_{2p}}(U) \leq N_{S_{2p}}(Z(G)) = N$, and thus $N_{S_{2p}}(U) = N_N(U)$. Let V denote a complement to $Z(G)$ in U, and let W be a Hall p'-group of N containing V (see [11, Hauptsatz VI.1.7]). Then W is a complement to $O_p(N)$ in N. As α centralizes ν, we have $\langle \nu, \alpha \rangle = \langle \nu \rangle \times \langle \alpha \rangle$, and thus $\langle \nu \rangle \times \langle \alpha \rangle$ is another complement to $O_p(N)$ in N. As all such complements are conjugate in N, there is $n \in N$ such that $W^n = \langle \nu \rangle \times \langle \alpha \rangle$ and $U^n = Z(G)V^n$. By replacing U with U^n, we may assume that $V \leq W = \langle \nu \rangle \times \langle \alpha \rangle$. In particular, W is abelian. It follows that $WZ(G)$ normalizes $U = VZ(G)$. As $U \nleq G$, there is an element $\nu^i\alpha^j \in V$ such that $\nu^i \neq 1$. Then $[\nu^i\alpha^j, \pi_1] \notin Z(G)$. In particular, U is not normal in N. As $WZ(G)$ has index p in N, we conclude that $N_N(U) = WZ(G)$, which proves our claim. □

Let n, d be positive integers, and let $\zeta \in S_n$ denote an n-cycle. Let us put

$$I_{n,d} := |\{\tau \in S_n \mid \tau^2 = 1, \tau\zeta^d = \zeta^d\tau\}| \quad (1)$$

and

$$I_n := I_{n,n}. \quad (2)$$

Thus, $I_{n,d}$ is one more than the number of involutions in $C_{S_n}(\zeta^d)$ and I_n is one more than the number of involutions in S_n. Notice that the definition of $I_{n,d}$ does not depend on the chosen n-cycle ζ, as all n-cycles are conjugate in S_n.

It is not difficult to derive a formula for $I_{n,d}$, where the formula for I_n is certainly well known. In the following result, $n \bmod 2 \in \{0, 1\}$ denotes the remainder of the division of n by 2.

Lemma 6.2 *Let n, d, e and f be positive integers such that $d \mid n$ and $\gcd(e, n/d) = 1$. Then, $I_{n,de} = I_{n,d}$ and $I_{n,f} = I_{n,\gcd(n,f)}$. Moreover, we have*

$$I_{n,d} = \sum_{k=0}^{\lfloor d/2 \rfloor} \frac{d!(n/d)^k(2-(n/d \bmod 2))^{d-2k}}{2^k k!(d-2k)!}.$$

In particular,

$$I_n = \sum_{k=0}^{\lfloor n/2 \rfloor} \frac{n!}{2^k k!(n-2k)!}.$$

Proof Let $\zeta \in S_n$ be an n-cycle. As e is relatively prime to n/d, we have that $\zeta^{de} = (\zeta^d)^e$ is the product of d cycles of length n/d. In particular, ζ^d and ζ^{de} are conjugate in S_n and thus $I_{n,de} = I_{n,d}$. Writing $f = de$ with $d = \gcd(n, f)$ and $e = f/\gcd(n, f)$, we obtain $I_{n,f} = I_{n,\gcd(n,f)}$, as $f/\gcd(n, f)$ and $n/\gcd(n, f)$ are relatively prime.

By definition, $I_{n,d}$ equals the number of elements $\tau \in C_{S_n}(\zeta^d)$ with $\tau^2 = 1$. The structure of $C_{S_n}(\zeta^d)$ is well known; it is a wreath product isomorphic to $C_{n/d} \wr S_d$, where $C_{n/d}$ denotes a cyclic group of order n/d. We view the elements of $C_{S_n}(\zeta^d)$ as $(d+1)$-tuples $(\mu; c_1, \ldots, c_d)$, where each c_i lies in one of the d cycles of ζ^d, and where $\mu \in S_d$ permutes the numbers $\{1, \ldots, d\}$. We have

$$(\mu; c_1, \ldots, c_d)^2 = (\mu^2; c_1 c_{1\mu^{-1}}, c_2 c_{2\mu^{-1}}, \ldots, c_d c_{d\mu^{-1}}).$$

Let $\tau := (\mu; c_1, \ldots, c_d) \in C_{S_n}(\zeta^d)$ satisfy $\tau^2 = 1$. Then, $\mu^2 = 1$ and $c_{i\mu} = c_i^{-1}$ for all $1 \le i \le d$. Suppose that μ is a product of exactly k transpositions for some $0 \le k \le \lfloor d/2 \rfloor$. Then, $c_j = c_i^{-1}$, if (i, j) is a transposition of μ, and $c_i^2 = 1$ if i is a fixed point of μ. This way, a fixed μ gives rise to $(n/d)^k(2-(n/d \bmod 2))^{d-2k}$ elements $\tau \in C_{S_n}(\zeta^d)$ with $\tau^2 = 1$. The centraliser of μ in S_d has order $2^k k!(d-2k)!$, yielding our formula for $I_{n,d}$. The one for I_n follows from this by putting $d = n$. □

Proposition 6.3 *There are exactly*

$$I_{p-1} - 1 + \frac{1}{p-1} \sum_{d=1}^{p-1} I_{p-1,d}$$

distinct isomorphism types of RCC loops with multiplication group G as in Case (a).

Proof Let (G, H, T) denote the envelope of an RCC loop of order $2p$ with G as in Case (a), i.e. G is isomorphic to the wreath product $C_p \wr C_2$. In this case, H is cyclic of order p. By numbering the right cosets of H in G from 1 to $2p$, we obtain an embedding $G \to S_{2p}$, and we identify G with its image in S_{2p} from now on. Let π_1, α

and π_2 be defined as in Lemma 6.1. We may choose the numbering of the cosets of H in G in such a way that $H = \langle \pi_2 \rangle$ and $G = \langle \pi_2, \alpha \rangle$. From Lemma 6.1(a) we obtain $Z(G) = \langle \pi_1\pi_2 \rangle$, $G' = \langle \pi_1^{-1}\pi_2 \rangle$ and $G = C_{S_{2p}}(Z(G))$. Also, $N := N_{S_{2p}}(G)$ equals $\langle \nu \rangle \ltimes G$ with ν as in Lemma 6.1. Observe that N normalizes H.

Let $\mathcal{T}$ denote the set of G-invariant transversals for $H\backslash G$ containing 1. Put $K := \langle \pi_1, \pi_2 \rangle = O_p(G)$ and let $\mathcal{T}_1$ denote the set of G-invariant transversals for $H\backslash K$ containing 1. Let $t \in G \setminus K$. Then, $|C_G(t)| = 2p$ and thus t lies in a conjugacy class of length p. As every conjugacy class of G lies in some coset of G', we find that $G't$ is the conjugacy class of G containing t. Hence, if $T \in \mathcal{T}$, we have $T = (K \cap T) \cup G't$ for some $t \in G \setminus K$, and $K \cap T \in \mathcal{T}_1$. Conversely, if $T_1 \in \mathcal{T}_1$, and if t is any element of $G \setminus K$, then $T_1 \cup G't \in \mathcal{T}$.

As $K = H \times H^\alpha$, we have $K = \cup_{0 \leq j \leq p-1} H\pi_1^j$. A transversal for $H\backslash K$ contains exactly one element of each coset $H\pi_1^j$, $0 \leq j \leq p-1$. As we insist that our transversals contain the trivial element, a transversal T_1 for $H\backslash K$ determines a map $\tau : \{1, \ldots, p-1\} \to \{0, 1, \ldots, p-1\}$ such that

$$T_1 = \{\pi_2^{j\tau}\pi_1^j \mid 1 \leq j \leq p-1\} \cup \{1\}. \quad (3)$$

Conjugating the element $\pi_2^{j\tau}\pi_1^j \in T_1 \setminus \{1\}$ by α, we obtain $\pi_1^{j\tau}\pi_2^j = \pi_2^j\pi_1^{j\tau}$. If T_1 is G-invariant, we must have, first, that $j\tau \neq 0$ and, second, that $\pi_2^j\pi_1^{j\tau} \in T_1 \setminus \{1\}$ for all $1 \leq j \leq p-1$. The latter condition implies that $\pi_2^{j\tau^2}\pi_1^{j\tau} = \pi_2^j\pi_1^{j\tau}$ for all $1 \leq j \leq p-1$, and thus $\tau^2 = 1$. In particular, τ is a permutation of order at most 2 of the set $\{1, \ldots, p-1\}$. Conversely, if τ is a permutation of the latter set with $\tau^2 = 1$, then T_1 defined by (3) lies in $\mathcal{T}_1$. In particular, $|\mathcal{T}_1| = I_{p-1}$. As the number of conjugacy classes of G in $G \setminus K$ equals p, we conclude from

$$\mathcal{T} = \{T_1 \cup G't \mid T_1 \in \mathcal{T}_1, t \in G \setminus K\},$$

that

$$|\mathcal{T}| = pI_{p-1}.$$

We next determine the number of N-orbits on $\mathcal{T}$. This is the same as the number of $\langle \nu \rangle$-orbits on $\mathcal{T}$. To compute this number, put

$$\mathcal{T}' := \{T_1 \cup G'\alpha \mid T_1 \in \mathcal{T}_1\} \subseteq \mathcal{T}.$$

Observe that $\mathcal{T}_1$ is $\langle \nu \rangle$-invariant, as ν normalizes H. In addition, ν centralizes α, and thus $\mathcal{T}'$ is $\langle \nu \rangle$-invariant as well. As $Z(G)$ is a set of representatives for the set of right cosets of G' in K, every conjugacy class of G contained in $G \setminus K$ is of the form $G'z\alpha$ for some $z \in Z(G)$. As $\langle \nu \rangle$ acts transitively on $Z(G) \setminus \{1\}$, we conclude that every orbit of $\langle \nu \rangle$ on $\mathcal{T} \setminus \mathcal{T}'$ has length $p-1$, and thus there are exactly I_{p-1} such orbits. We are thus left with the determination of the number of $\langle \nu \rangle$-orbits on $\mathcal{T}'$, which is the same as the number of $\langle \nu \rangle$-orbits on $\mathcal{T}_1$. By the Burnside–Cauchy–Frobenius lemma, the latter number equals

$$\frac{1}{p-1}\sum_{d=1}^{p-1}\chi_d,$$

where χ_d is the number of fixed points of ν^d on $\mathcal{T}_1$. The action of $\langle\nu\rangle$ on K determines a $(p-1)$-cycle ζ on the set $\{1,\ldots,p-1\}$ such that $\nu^{-1}\pi_i^j\nu = \pi_i^{j\zeta}$ for $i=1,2$ and all $1\le j\le p-1$. Now let $T_1\in\mathcal{T}_1$ be given by (3) with respect to $\tau\in S_{p-1}$ with $\tau^2=1$. Then, T_1 is fixed by ν^d, if and only if ζ^d centralizes τ. Thus $\chi_d = I_{p-1,d}$.

It remains to determine those N-orbits on $\mathcal{T}$ containing transversals that generate G. Let $T\in\mathcal{T}$ such that $\langle T\rangle\ne G$. Then, $\langle T\rangle$ is a normal subgroup of G of index p. Thus, $G'\le\langle T\rangle$ and $T=G'\cup G'z\alpha$ for some $z\in Z(G)$. Since $\langle T\rangle\ne G$, we must have $z=1$, i.e. $\langle T\rangle = T = G'\cup G'\alpha$. As this is N-invariant, our result follows. □

Proposition 6.4 *Write $p-1=2^n r$ with positive integers n and r and with r odd. Then, there are exactly $p-r-1$ distinct isomorphism types of RCC loops with multiplication group G as in* Case (b), *and there are exactly r isomorphism types of RCC loops with multiplication group G as in* Case (c).

Proof Let (G,H,T) denote the envelope of an RCC loop of order $2p$ with G as in Case (b) or (c). If $H=1$, then $T=G$ is a group of order $2p$, which is non-abelian in Case (b), and cyclic in Case (c). In each case, we obtain a unique isomorphism class of RCC loops.

Thus, let us assume that $H\ne 1$ in the following. As in the proof of Proposition 6.3, we identify G with its image in S_{2p} through an embedding obtained by numbering the right cosets of H in G from 1 to $2p$. Put $P := O_p(G)$, the unique Sylow p-subgroup of G. Let π_1, α and π_2 be defined as in Lemma 6.1. We may choose the numbering of the cosets of H in G in such a way that $P=\langle\pi_1\pi_2\rangle$, and that $Z(G)=\langle\alpha\rangle$ in Case (c). Put $N := N_{S_{2p}}(G)$. We now apply Lemma 6.1(b) with our G taking the role of U of that lemma. As $H\ne 1$, we have $G\not\le\langle\pi_2,\alpha\rangle$, and thus, replacing G by a suitable conjugate within $N_{S_{2p}}(P)$, we find that $N = A\times\langle\alpha\rangle$, with $A\cong \mathrm{Aff}(1,p)$. We have $A = L\ltimes P$, with L cyclic of order $p-1$.

Assume that G is as in Case (b). Then, $G\cap L$ is a complement to P in G. As all such complements are conjugate in G by Schur's theorem (see [11, Satz I.17.5]), we may assume that $H\le L$. In particular, $G\le A$, and H is N-invariant. Let T be a G-invariant transversal for $H\backslash G$. Then $P\subseteq T$ by Theorem 5.13. Let $\tau\in T\setminus P$. Then $|C_G(\tau)| = 2|H|$ and thus $T\setminus P$ consists of the G-conjugacy class containing τ. If, moreover, $G=\langle P,\tau\rangle$, we have $2p|H| = |G| = p|\tau|$ and τ has even order larger than 2. Every element τ' which is conjugate to τ in A gives rise to an isomorphic loop with multiplication group $\langle P,\tau'\rangle$, as in Case (b). It follows that the isomorphism types of RCC loops with a multiplication group as in Case (b) equals the number of A-conjugacy classes of elements of A of even order larger than 2. As A has $(p-r-2)p$ such elements, the result follows.

Assuming now that G is as in Case (c), we have $G = K\times\langle\alpha\rangle$, with $K = O_{2'}(G)$, and thus $K\trianglelefteq N$. In turn, $K\le A$ as every Sylow subgroup of K is conjugate to a subgroup of A. Again, H is N-invariant. Let T be a G-invariant transversal for $H\backslash G$. As in Case (b), we have $T = P \cup C$, where C is a G-conjugacy class of an element $\tau \in G \setminus K$. Every element τ' in the A-conjugacy class containing τ gives rise to an isomorphic loop with multiplication group $\langle P, \tau' \rangle$. Now $\tau = \tau_1\alpha$ for some $\tau_1 \in K$. It follows that the isomorphism types of RCC loops with a multiplication group as in Case (c) equals the number of A-conjugacy classes of elements of A of odd order different from p. All these elements lie in the unique subgroup of A of order pr, and thus there are $(r-1)p$ non-trivial such elements. As the trivial element yields a group, the result follows. □

We summarize our results in the following theorem.

Theorem 6.5 *Let p be a prime. Then, the number of isomorphism types of RCC loops of order $2p$ (including groups) equals*

$$p - 2 + I_{p-1} + \frac{1}{p-1}\sum_{d=1}^{p-1} I_{p-1,d}. \tag{4}$$

Proof Every loop of order 4 is a group. As $I_{1,1} = I_1 = 1$, formula (4) holds for $p = 2$. For odd p it follows from Propositions 6.3 and 6.4, as the cases in Theorem 5.13 are disjoint. □

The table below contains the numbers obtained by evaluating formula (4) for small values of p. These numbers have also been obtained for $p \leq 13$ in the PhD thesis of the first author [1] by different methods.

p	2	3	5	7	11	13	17	19
(4)	2	5	18	99	10 489	151 973	49 096 721	1 052 729 657

One of the referees has kindly pointed out that formula (4) evaluates to an integer, even if p is not a prime (and larger than 1). This follows from the fact that for general positive integers n, d, the number $I_{n,d}$ equals the number of fixed points of the element ζ^d on the set $\{\tau \in S_n \mid \tau^2 = 1\}$, where the n-cycle ζ acts by conjugation. Thus, by the Burnside–Cauchy–Frobenius lemma, the number of orbits of $\langle \zeta \rangle$ on $\{\tau \in S_n \mid \tau^2 = 1\}$ equals $1/n \sum_{d=1}^{n} I_{n,d}$, so that this number is an integer.

7 A Series of Examples

According to Theorem 5.13, the right multiplication group of an RCC loop of order $2p$, where p is an odd prime, is soluble. This is no longer the case for right multiplication groups of RCC loops of order pq, where p and q are distinct primes. An example is given in [1, Table B.7] of an RCC loop of order 15 with right multiplication group isomorphic to GL(2, 4). This fits into an infinite series of examples.

Proposition 7.1 *Let q be a power of a prime with $q > 2$. Then, there is an RCC loop of of order $q^2 - 1$ and right multiplication group isomorphic to* $\mathrm{GL}(2, q)$.

Proof Let $G := \mathrm{GL}(2, q)$, acting from the right on $\mathbb{F}_q^{1\times 2}$, and let

$$H := \left\{ \begin{pmatrix} \alpha & 0 \\ \beta & 1 \end{pmatrix} \mid \alpha \in \mathbb{F}_q^*, \beta \in \mathbb{F}_q \right\}.$$

Let $Z := Z(G)$ denote the set of scalar matrices in G and let C be a G-conjugacy class of elements of order $q^2 - 1$, i.e. the elements of C are Singer cycles. Then, $|C_G(t)| = q^2 - 1$ for all $t \in C$; in particular $|C| = q(q - 1)$. Now, put

$$T := C \cup Z.$$

We claim that T is a G-invariant transversal for $H\backslash G$. Clearly, T is G-invariant and $|T| = q^2 - 1 = |G : H|$. Let $t, t' \in C$. We have to show that $t't^{-1} \in H$ if and only if $t = t'$. To see this, first observe that $|C_G(t)||H| = |G|$ and that $C_G(t) \cap H = 1$, as $|C_G(t) \cap H|$ divides $\gcd(|C_G(t)|, |H|) = q - 1$, and the only elements in $C_G(t)$ of order dividing $q - 1$ are the elements of Z. We conclude that $G = C_G(t)H$. It follows that there is $h \in H$ with $t' = h^{-1}th$. Put $h' := t't^{-1} = h^{-1}tht^{-1}$. Thus,

$$t^{-1}hh' = ht^{-1}. \tag{5}$$

Now, assume that $h' \in H$. As $\det(h') = \det(t't^{-1}) = 1$, we have

$$h' = \begin{pmatrix} 1 & 0 \\ \gamma & 1 \end{pmatrix}$$

for some $\gamma \in \mathbb{F}_q$. Let

$$h = \begin{pmatrix} \alpha & 0 \\ \beta & 1 \end{pmatrix}$$

with $\alpha \in \mathbb{F}_q^*$ and $\beta \in \mathbb{F}_q$, and let

$$t^{-1} = \begin{pmatrix} a & b \\ c & d \end{pmatrix}$$

with $a, b, c, d \in \mathbb{F}_q$. Then

$$t^{-1}hh' = \begin{pmatrix} * & b \\ * & d \end{pmatrix},$$

and

$$ht^{-1} = \begin{pmatrix} * & \alpha b \\ * & \beta b + d \end{pmatrix},$$

where we do not need to specify the entries in the first columns of $t^{-1}hh'$, respectively ht^{-1}. As t acts irreducibly on the natural vector space $\mathbb{F}_q^{1\times 2}$ for G, we conclude that $b \neq 0$. Equation (5) yields $\alpha = 1$ and $\beta = 0$, i.e. $h = 1$, and thus $t = t'$. If $z, z' \in Z$, then $z'z^{-1} \in H$ if and only if $z = z'$. Now, let $z \in Z$ and $t \in C$ and assume that $tz^{-1} \in H$. Then, $t \in HZ$; but $|HZ| = q(q-1)^2$, whereas $|t| = q^2 - 1 \nmid q(q-1)^2$, a contradiction.

Finally, it is easy to check that $\langle T \rangle = G$, by a direct computation if $q = 3$, and using the fact that G/Z is almost simple if $q \neq 3$. This completes the proof. □

Acknowledgements We thank Alice Niemeyer for her support and her interest in this work. We also thank Barbara Baumeister for introducing us to the fascinating topic of invariant transversals. Finally, we are very much indebted to the anonymous referees for several suggestions which improved the exposition of this paper, and also for drawing our attention to related work.

References

1. K. Artic, On right conjugacy closed loops and right conjugacy closed loop folders, Dissertation, RWTH Aachen University, 2017
2. M. Aschbacher, On Bol loops of exponent 2. J. Algebra **288**, 99–136 (2005)
3. R. Baer, Nets and groups. Trans. Am. Math. Soc. **46**, 110–141 (1939)
4. R.P. Burn, Finite Bol loops. Math. Proc. Camb. Philos. Soc. **84**, 377–385 (1978)
5. P. Csőrgő, M. Niemenmaa, On connected transversals to nonabelian subgroups. Eur. J. Comb. **23**, 179–185 (2002)
6. P. Csőrgő, A. Drápal, Left conjugacy closed loops of nilpotency class two. Results Math. **47**, 242–265 (2005)
7. D. Daly, P. Vojtěchovský, Enumeration of nilpotent loops via cohomology. J. Algebra **322**, 4080–4098 (2009)
8. J.D. Dixon, B. Mortimer, *Permutation Groups* (Springer, Berlin, 1996)
9. A. Drápal, On multiplication groups of left conjugacy closed loops. Comment. Math. Univ. Carolin. **45**, 223–236 (2004)
10. B. Huppert, N. Blackburn, *Finite Groups II* (Springer, Berlin, 1982)
11. B. Huppert, *Endliche Gruppen I* (Springer, Berlin, 1967)
12. B. Huppert, Singerzyklen in klassischen Gruppen. Math. Z. **117**, 141–150 (1970)
13. P. Kleidman, M. Liebeck, *in The Subgroup Structure of the Finite Classical Groups, London mathematical society, Lecture note series*, vol. 129 (Cambridge University Press, Cambridge, 1990)
14. K. Kunen, The structure of conjugacy closed loops. Trans. Am. Math. Soc. **352**, 2889–2911 (2000)
15. C.H. Li, Á. Seress, The primitive permutation groups of squarefree degree. Bull. Lond. Math. Soc. **35**, 635–644 (2003)
16. G. Nagy, P. Vojtěchovský, Loops: Computing with quasigroups and loops in GAP, Version 3.0.0, *Refereed GAP package* (2015). http://www.math.du.edu/loops
17. P.A.B. Pleasants, The number of prime factors of binomial coefficients. J. Number Theory **15**, 203–225 (1982)
18. A. Stein, A conjugacy class as a transversal in a finite group. J. Algebra **239**, 365–390 (2001)
19. D.E. Taylor, *The Geometry of the Classical Groups* (Heldermann, Berlin, 1992)

20. The GAP Group, GAP – Groups, Algorithms, and Programming, Version 4.8.7 (2017). http://www.gap-system.org
21. A.V. Vasilyev, Minimal permutation representations of finite simple exceptional groups of types G_2 and F_4. Algebra Log. **35**, 371–383 (1996)
22. A.V. Vasilyev, Minimal permutation representations of finite simple exceptional groups of types E_6, E_7 and E_8. Algebra Log. **36**, 302–310 (1997)
23. A.V. Vasilyev, Minimal permutation representations of finite simple exceptional groups of twisted type. Algebra Log. **37**, 9–20 (1998)

Compatible Actions and Non-abelian Tensor Products

Valeriy G. Bardakov and Mikhail V. Neshchadim

2010 Mathematics Subject Classification Primary 20E22 · Secondary 20F18 20F28

1 Introduction

R. Brown and J.-L. Loday [1, 2] introduced the non-abelian tensor product $G \otimes H$ for a pair of groups G and H following the works of C. Miller [6], and A. S.-T. Lue [5]. The investigation of the non-abelian tensor product from a group theoretical point of view started with a paper by R. Brown, D. L. Johnson, and E. F. Robertson [3].

The non-abelian tensor product $G \otimes H$ depends not only on the groups G and H but also on the action of G on H and on the action of H on G. Usually, people consider compatible actions (see the definition in Sect. 2). In the present paper, we study the following question: What actions are compatible?

V. G. Bardakov (✉)
Sobolev Institute of Mathematics, Novosibirsk 630090, Russia
e-mail: bardakov@math.nsc.ru

V. G. Bardakov
Novosibirsk State University, Novosibirsk 630090, Russia

V. G. Bardakov
Novosibirsk State Agrarian University, Dobrolyubova Street, 160,
Novosibirsk 630039, Russia

M. V. Neshchadim
Sobolev Institute of Mathematics and Novosibirsk State University,
Novosibirsk 630090, Russia
e-mail: neshch@math.nsc.ru

N. S. N. Sastry and M. K. Yadav (eds.), *Group Theory and Computation*,
Indian Statistical Institute Series, https://doi.org/10.1007/978-981-13-2047-7_2

The paper is organized as follows. In Sect. 2, we recall a definition of non-abelian tensor product, formulate some its properties and give an answer on a question of V. Thomas, proving that there are nilpotent group G and some group H such that for $G \otimes H$ the derivative group $[G, H]$ is equal to G. In the Sect. 3, we study the following question: Let a group H act on a group G by automorphisms, is it possible to define an action of G on H such that this pair of actions is compatible? Some necessary conditions for compatibility of actions will be given and in some cases will prove a formula for the second action if the first one is given. In the Sect. 4, we construct pairs compatible actions for arbitrary groups and for nilpotent groups of class 2, that give a particular answer on the question from Sect. 3. In Sect. 5 we study groups of the form $G \otimes \mathbb{Z}_2$ and describe compatible actions.

2 Preliminaries

In this article, we will use the following notations. For elements x, y in a group G, the conjugation of x by y is $x^y = y^{-1}xy$; and the commutator of x and y is $[x, y] = x^{-1}x^y = x^{-1}y^{-1}xy$. We write G' for the derived subgroup of G, i.e., $G' = [G, G]$; G^{ab} for the abelianized group G/G'; the second hypercenter $\zeta_2 G$ of G is the subgroup of G such that

$$\zeta_2 G/\zeta_1 G = \zeta_1(G/\zeta_1 G),$$

where $\zeta_1 G = Z(G)$ is the center of a group G.

Recall the definition of the non-abelian tensor product $G \otimes H$ of groups G and H (see [1, 2]). It is defined for a pair of groups G and H where each one acts on the other (on right)

$$G \times H \longrightarrow G, \ \ (g, h) \mapsto g^h; \ \ \ H \times G \longrightarrow H, \ \ (h, g) \mapsto h^g$$

and on itself by conjugation, in such a way that for all $g, g_1 \in G$ and $h, h_1 \in H$,

$$g^{(h^{g_1})} = \left(\left(g^{g_1^{-1}}\right)^h\right)^{g_1} \ \text{ and } \ h^{(g^{h_1})} = \left(\left(h^{h_1^{-1}}\right)^g\right)^{h_1}.$$

In this situation, we say that G and H act *compatibly* on each other. The *non-abelian tensor product* $G \otimes H$ is the group generated by all symbols $g \otimes h$, $g \in G$, $h \in H$, subject to the relations

$$gg_1 \otimes h = (g^{g_1} \otimes h^{g_1})(g_1 \otimes h) \text{ and } \ g \otimes hh_1 = (g \otimes h_1)(g^{h_1} \otimes h^{h_1})$$

for all $g, g_1 \in G$, $h, h_1 \in H$.

In particular, as the conjugation action of a group G on itself is compatible, then the tensor square $G \otimes G$ of a group G may always be defined. Also, the tensor product $G \otimes H$ is defined if G and H are two normal subgroups of some group M and actions are conjugations in M.

The following proposition is well known. We give a proof only for fullness.

Proposition 2.1 *(1) Let G and H be abelian groups. Independently of the action of G on H and H on G, the group $G \otimes H$ is abelian.*

(2) (See [2, Proposition 2.4]) Let G and H be arbitrary groups. If the actions of G on H and H on G are trivial, then the group $G \otimes H \cong G^{ab} \otimes_{\mathbb{Z}} H^{ab}$ is the abelian tensor product.

Proof (1) We have the equality

$$(g \otimes h)^{g_1 \otimes h_1} = g^{[g_1, h_1]} \otimes h^{[g_1, h_1]},$$

where $g^{[g_1, h_1]}$ is the action of the commutator $[g_1, h_1] \in G$ by conjugation on g, but G is abelian and $g^{[g_1, h_1]} = g$. Analogously, $h^{[g_1, h_1]} = h$. Hence, $G \otimes H$ is abelian.

(2) From the previous formula and triviality actions, we have

$$g^{[g_1, h_1]} = g^{g_1^{-1} h_1^{-1} g_1 h_1} = \left(g^{g_1^{-1}}\right)^{h_1^{-1} g_1 h_1} = \left(g^{g_1^{-1}}\right)^{g_1 h_1} = \left(g^{g_1^{-1}}\right)^{g_1 h_1} = g^{h_1} = g.$$

Analogously, $h^{[g_1, h_1]} = h$. Hence, $G \otimes H$ is abelian. □

A subgroup of G called the *derivative subgroup* of G by H was introduce in [7]. It is defined as

$$D_H(G) = [G, H] = \langle g^{-1} g^h \mid g \in G, h \in H \rangle.$$

Also remind some approach to description of non-abelian tensor product (see [4]).

The map $\kappa : G \otimes H \longrightarrow D_H(G)$ defined by $\kappa(g \otimes h) = g^{-1} g^h$ is a homomorphism, its kernel $A = \ker(\kappa)$ is the central subgroup of $G \otimes H$ and G acts on $G \otimes H$ by the rule $(g \otimes h)^x = g^x \otimes h^x$, $x \in G$, i.e., there exists the short exact sequence

$$1 \longrightarrow A \longrightarrow G \otimes H \longrightarrow D_H(G) \longrightarrow 1.$$

In this case, A can be viewed as trivial $\mathbb{Z}[D_H(G)]$-module via conjugation in $G \otimes H$.

The following proposition gives an answer on the following question: Is there non-abelian tensor product $G \otimes H$ such that $[G, H] = G$? which of V. Thomas formulated in a private communication to the authors.

Proposition 2.2 *Let $G = F_n / \gamma_k F_n$, $k \geq 2$, be a free nilpotent group of rank $n \geq 2$ and $H = \mathrm{Aut}(G)$ is its automorphism group. Then, $D_H(G) = [G, H] = G$.*

Proof Let F_n be a free group of rank $n \geq 2$ with the basis $x_1, \ldots, x_n$, $G = F_n / \gamma_k F_n$ be a free nilpotent group of class $k - 1$ for $k \geq 2$. Let G act trivially on H and elements of H act by automorphisms on G. It is easy to see that these actions are compatible.

Let us show that in this case $[G, H] = G$. To do it, let us prove that x_1 lies in $[G, H]$. Take $\varphi_1 \in H = \mathrm{Aut}(G)$, which acts on the generators of G by the rules:

$$x_1^{\varphi_1} = x_1, \ \ x_2^{\varphi_1} = x_2 x_1, \ \ x_3^{\varphi_1} = x_3, \ldots, x_n^{\varphi_1} = x_n.$$

Then,

$$x_1^{-1} x_1^{\varphi_1} = 1, \ \ x_2^{-1} x_2^{\varphi_1} = x_1, \ \ x_3^{-1} x_3^{\varphi_1} = 1, \ldots, x_n^{-1} x_n^{\varphi_1} = 1.$$

Hence, the generator x_1 lies in $[G, H]$. Analogously, $x_2, x_3, \ldots, x_n$ lie in $[G, H]$. This completes the proof. □

3 What Actions Are Compatible?

In this section, we study

Question 1 *Let a group H acts on a group G by automorphisms. Is it possible to define a non-abelian tensor product $G \otimes H$ with compatible actions of G on H and H on G?*

Consider some examples.

Example 3.1 Let us take $G = \{1, a, a^2\} \cong \mathbb{Z}_3$, $H = \{1, b, b^2\} \cong \mathbb{Z}_3$. In dependence on actions we have three cases.

(1) If the action of H on G and the action of G on H are trivial, then by the second part of Proposition 2.1 $G \otimes H = \mathbb{Z}_3 \otimes_{\mathbb{Z}} \mathbb{Z}_3 \cong \mathbb{Z}_3$ is abelian tensor product.

(2) Let H acts nontrivially on G, i.e., $a^b = a^2$ and the action G on H is trivial. It is not difficult to check that G and H act compatibly on each other. To find $D_H(G) = [G, H]$, we calculate

$$[a, b] = a^{-1} a^b = a^2 a^2 = a.$$

Hence, $D_H(G) = G$. But $D_G(H) = 1$.

By the definition, $G \otimes H$ is generated by elements

$$a \otimes b, a^2 \otimes b, a \otimes b^2, a^2 \otimes b^2.$$

Using the defining relations

$$g g_1 \otimes h = (g^{g_1} \otimes h^{g_1})(g_1 \otimes h), \quad g \otimes h h_1 = (g \otimes h_1)(g^{h_1} \otimes h^{h_1}),$$

we find

$$a^2 \otimes b = (a^a \otimes b^a)(a \otimes b) = (a \otimes b)^2, \;\; a \otimes b^2 = (a \otimes b)(a^b \otimes b^b)$$
$$= (a \otimes b)(a^2 \otimes b) = (a \otimes b)^3.$$

On the other side,

$$1 = a^2 a \otimes b = (a^2 \otimes b^a)(a \otimes b) = (a \otimes b)^3.$$

Hence,

$$a \otimes b^2 = a^2 \otimes b^2 = 1$$

and in this case, we have the same result: $\mathbb{Z}_3 \otimes \mathbb{Z}_3 = \mathbb{Z}_3$.

(3) Let H act nontrivially on G, i.e., $a^b = a^2$ and G act nontrivially on H. In this case, G and H acts non-compatibly on each other. Indeed,

$$a^{(b^a)} = a^{b^2} = (a^2)^b = a,$$

but

$$\left(\left(a^{a^{-1}}\right)^b\right)^a = (a^b)^a = (a^2)^2 = a^2.$$

Hence, the equality

$$a^{(b^a)} = \left(\left(a^{a^{-1}}\right)^b\right)^a$$

does not hold.

Let G, H be some groups. Actions of G on H and H on G are defined by homomorphisms

$$\beta : G \to \mathrm{Aut}(H), \quad \alpha : H \to \mathrm{Aut}(G),$$

and by definition

$$g^h = g^{\alpha(h)}, \quad h^g = h^{\beta(g)}, \quad g \in G, h \in H.$$

The actions (α, β) are compatible, if

$$g^{\alpha(h^{\beta(g_1)})} = \left(\left(g^{g_1^{-1}}\right)^{\alpha(h)}\right)^{g_1}$$

and

$$h^{\beta(g^{\alpha(h_1)})} = \left(\left(h^{h_1^{-1}}\right)^{\beta(g)}\right)^{h_1}$$

for all $g, g_1 \in G, h, h_1 \in H$. In this case, we will say that the pair (α, β) *is compatible*.

Rewrite these equalities in the form:

$$\alpha\left(h^{\beta(g_1)}\right) = \widehat{g_1}^{-1}\alpha(h)\widehat{g_1} \tag{1}$$

and

$$\beta\left(g^{\alpha(h_1)}\right) = \widehat{h_1}^{-1}\beta(g)\widehat{h_1}, \tag{2}$$

where $\widehat{g}$ is the inner automorphism of G which is induced by conjugation of g, i.e.,

$$\widehat{g} : g_1 \mapsto g^{-1}g_1g, \quad g, g_1 \in G,$$

and analogously, $\widehat{h}$ is the inner automorphism of H which is induced by the conjugation of h, i.e.,

$$\widehat{h} : h_1 \mapsto h^{-1}h_1h, \quad h, h_1 \in H.$$

Theorem 3.2 *(1) If the pair (α, β) defines compatible actions of H on G and G on H, then the following inclusions hold*

$$N_{\mathrm{Aut}(G)}(\alpha(H)) \geq \mathrm{Inn}(G), \quad N_{\mathrm{Aut}(H)}(\beta(G)) \geq \mathrm{Inn}(H).$$

Here, $\mathrm{Inn}(G)$ *and* $\mathrm{Inn}(H)$ *are the subgroups of inner automorphisms.*

(2) If $\alpha : H \to \mathrm{Aut}(G)$ is an embedding and $N_{\mathrm{Aut}(G)}(\alpha(H)) \geq \mathrm{Inn}(G)$, then defining $\beta : G \to \mathrm{Aut}(H)$ by the formula

$$\beta(g) : h \mapsto \alpha^{-1}\left(\widehat{g}^{-1}\alpha(h)\widehat{g}\right), \quad h \in H,$$

we get the compatible actions (α, β).

Proof The first claim immediately follows from the relations (1), (2).

To prove the second claim, it is enough to check (2), or that is equivalent, the equality

$$h^{\beta(g^{\alpha(h_1)})} = \left(\left(h^{h_1^{-1}}\right)^{\beta(g)}\right)^{h_1}. \tag{3}$$

Using the definition β, rewrite the left side of (3):

$$h^{\beta(g^{\alpha(h_1)})} = \alpha^{-1}\left(\widehat{g^{\alpha(h_1)}}^{-1}\alpha(h)\widehat{g^{\alpha(h_1)}}\right). \tag{4}$$

Rewrite the right side of (3):

$$\left(\left(h^{h_1^{-1}}\right)^{\beta(g)}\right)^{h_1} = h_1^{-1}(h_1hh_1^{-1})^{\beta(g)}h_1 = h_1^{-1}\alpha^{-1}(\widehat{g}^{-1}\alpha(h_1hh_1^{-1})\widehat{g})h_1. \tag{5}$$

From (4) and (5),

$$\alpha^{-1}\left(\widehat{g^{\alpha(h_1)}}^{-1}\alpha(h)\widehat{g^{\alpha(h_1)}}\right) = h_1^{-1}\alpha^{-1}(\widehat{g}^{-1}\alpha(h_1hh_1^{-1})\widehat{g})h_1.$$

Using the homomorphism α

$$\widehat{g^{\alpha(h_1)}}^{-1}\alpha(h)\widehat{g^{\alpha(h_1)}} = \alpha\left(h_1^{-1}\alpha^{-1}(\widehat{g}^{-1}\alpha(h_1hh_1^{-1})\widehat{g})h_1\right) =$$
$$= \alpha(h_1)^{-1}\widehat{g}^{-1}\alpha(h_1hh_1^{-1})\widehat{g}\alpha(h_1) =$$
$$= \alpha(h_1)^{-1}\widehat{g}^{-1}\alpha(h_1)\alpha(h)\alpha(h_1)^{-1})\widehat{g}\alpha(h_1) = \widehat{g^{\alpha(h_1)}}^{-1}\alpha(h)\widehat{g^{\alpha(h_1)}}.$$

In the last equality, we used the formula

$$\alpha(h_1)^{-1}\widehat{g}\alpha(h_1) = \widehat{g^{\alpha(h_1)}}.$$

Hence, the equality (3) holds. □

Conjecture 1 *Are the inclusions*

$$N_{\mathrm{Aut}(G)}(\alpha(H)) \geq \mathrm{Inn}(G), \quad N_{\mathrm{Aut}(H)}(\beta(G)) \geq \mathrm{Inn}(H)$$

sufficient for compatibility of the pair (α, β)*?*

4 Compatible Actions for Nilpotent Groups

At first, let us give the following definition.

Definition 4.1 Let G and H be groups and $G_1 \trianglelefteq G$, $H_1 \trianglelefteq H$ are their normal subgroups. We will say that G is *comparable with* H *with respect to the pair* (G_1, H_1), if there are homomorphisms

$$\varphi : G \longrightarrow H, \quad \psi : H \longrightarrow G,$$

such that

$$x \equiv \psi\varphi(x)(\mathrm{mod}\, G_1), \quad y \equiv \varphi\psi(y)(\mathrm{mod}\, H_1)$$

for all $x \in G$, $y \in H$, i.e.,

$$x^{-1} \cdot \psi\varphi(x) \in G_1, \quad y^{-1} \cdot \varphi\psi(y) \in H_1.$$

Note that if $G_1 = 1$, $H_1 = 1$, then φ, ψ are mutually inverse isomorphisms.
The following theorem holds.

Theorem 4.2 *Let G, H be groups and there exist homomorphisms*

$$\varphi : G \longrightarrow H, \quad \psi : H \longrightarrow G,$$

such that

$$x \equiv \psi\varphi(x)(\mathrm{mod}\, \zeta_2 G), \quad y \equiv \varphi\psi(y)(\mathrm{mod}\, \zeta_2 H)$$

for all $x \in G$, $y \in H$. Then, the action of G on H and the action of H on G by the rules

$$x^y = \psi(y)^{-1} x \psi(y), \quad y^x = \varphi(x)^{-1} y \varphi(x), \quad x \in G, \ y \in H,$$

are compatible, i.e., the following equalities hold

$$x^{(y^{x_1})} = ((x^{x_1^{-1}})^y)^{x_1}, \quad y^{(x^{y_1})} = ((y^{y_1^{-1}})^x)^{y_1}, \quad x, x_1 \in G, \ y, y_1 \in H.$$

Proof Let us prove that the following relation holds:

$$x^{(y^{x_1})} = ((x^{x_1^{-1}})^y)^{x_1}.$$

For this denote, the left-hand side of this relation by L and transform it

$$\begin{aligned} L = x^{(y^{x_1})} &= x^{\varphi(x_1)^{-1} y \varphi(x_1)} = \psi(\varphi(x_1)^{-1} y^{-1} \varphi(x_1)) x \psi(\varphi(x_1)^{-1} y \varphi(x_1)) = \\ &= (\psi\varphi(x_1))^{-1} \psi(y)^{-1} (\psi\varphi(x_1)) x (\psi\varphi(x_1))^{-1} \psi(y) (\psi\varphi(x_1)) = \\ &= (c(x_1)^{-1} x_1^{-1} \psi(y)^{-1} x_1 c(x_1)) x (c(x_1)^{-1} x_1^{-1} \psi(y) x_1 c(x_1)). \end{aligned}$$

Here, $\psi\varphi(x_1) = x_1 c(x_1)$, $c(x_1) \in \zeta_2 G$. Since $c(x_1) \in \zeta_2 G$, then the commutator $[x_1^{-1}\psi(y)x_1, c(x_1)]$ lies in the center of G. Hence,

$$L = x^{x_1^{-1}\psi(y)x_1}.$$

Denote the right-hand side of this relation by R and transform it

$$R = ((x^{x_1^{-1}})^y)^{x_1} = ((x^{x_1^{-1}})^{\psi(y)})^{x_1} = x^{x_1^{-1}\psi(y)x_1}.$$

We see that $L = R$, i.e., the first relation from the definition of compatible action holds. The checking of the second relation is the similar. □

From this theorem, we have a particular answer on Question 1 for nilpotent groups of class 2.

Corollary 4.3 *If G, H are nilpotent groups of class 2, then any pair of homomorphisms*

$$\varphi : G \longrightarrow H, \quad \psi : H \longrightarrow G$$

define the compatible action.

Problem 1 Let G and H be free nilpotent groups of class 2. By Corollary 4.3, any pair of homomorphisms (φ, ψ), where $\varphi \in \mathrm{Hom}(G, H)$, $\psi \in \mathrm{Hom}(H, G)$ defines a tensor product $M(\varphi, \psi) = G \otimes H$. Give a classification of the groups $M(\varphi, \psi)$.

Note that for arbitrary groups, Corollary 4.3 does not hold. Indeed, let $G = \langle x_1, x_2 \rangle$, $H = \langle y_1, y_2 \rangle$ be free groups of rank 2. Define the homomorphisms

$$\varphi : G \longrightarrow H, \quad \psi : H \longrightarrow G$$

by the rules

$$\varphi(x_1) = y_1, \ \varphi(x_2) = y_2, \quad \psi(y_1) = \psi(y_2) = 1.$$

Then

$$y_2^{x_1} = y_2^{\varphi(x_1)} = y_2^{y_1} \neq y_2,$$

i.e., the conditions of compatible actions does not hold.

5 Tensor Products $G \otimes \mathbb{Z}_2$

Note that the group $\mathrm{Aut}(\mathbb{Z}_2)$ is trivial and hence, any group G acts on $\mathbb{Z}_2$ only trivially.

This section is devoted to answer the following question.

Question 2 *Let G be a group and $\psi \in \mathrm{Aut}(G)$ be an automorphism of order 2. Let $\mathbb{Z}_2 = \langle \varphi \rangle$ and $\alpha : \mathbb{Z}_2 \longrightarrow \mathrm{Aut}(G)$ such that $\alpha(\varphi) = \psi$. Under what conditions the pare $(\alpha, 1)$ is compatible?*

If $\psi \in \mathrm{Aut}(G)$ is trivial automorphism, then by the second part of Proposition 2.1 $G \otimes \mathbb{Z}_2 = G^{ab} \otimes_{\mathbb{Z}} \mathbb{Z}_2$ is an abelian tensor product. In the general case, we have

Proposition 5.1 *Let*

(1) G be a group,

(2) $\mathbb{Z}_2 = \langle \varphi \rangle$ be a cyclic group of order two with the generator φ,

(3) $\alpha : \mathbb{Z}_2 \longrightarrow \mathrm{Aut}(G)$ be a homomorphism, $\beta = 1 : G \to \mathrm{Aut}(\mathbb{Z}_2)$ be the trivial homomorphism,

Then, the pair of actions (α, β) is compatible if and only if for any $g \in G$ holds

$$g^{\alpha(\varphi)} = gc(g),$$

where $c(g)$ is a central element of G such that $c(g)^{\alpha(\varphi)} = c(g)^{-1}$. In particular, if the center of G is trivial, then $G \otimes \mathbb{Z}_2 = G^{ab} \otimes_{\mathbb{Z}} \mathbb{Z}_2$.

Proof Since $\mathrm{Inn}(G)$ normalizes $\alpha(\mathbb{Z}_2)$, then for every $g \in G$ holds

$$\widehat{g}^{-1}\alpha(\varphi)\widehat{g} = \alpha(\varphi).$$

Using this equality for arbitrary element $x \in G$, we get

$$g^{-1}g^{\alpha(\varphi)}x^{\alpha(\varphi)}(g^{-1}g^{\alpha(\varphi)})^{-1} = x^{\alpha(\varphi)}.$$

Since $x^{\alpha(\varphi)}$ is an arbitrary element of G, then $c(g)$ is a central element of G. Applying $\alpha(\varphi)$ to the equality $g^{\alpha(\varphi)} = gc(g)$, we have

$$g = g^{\alpha(\varphi)^2} = g^{\alpha(\varphi)}c(g)^{\alpha(\varphi)} = gc(g)c(g)^{\alpha(\varphi)},$$

that is $c(g)^{\alpha(\varphi)} = c(g)^{-1}$. □

For an arbitrary abelian group A, we know that $A \otimes_{\mathbb{Z}} \mathbb{Z} = A$. The following proposition is some analog of this property for non-abelian tensor product.

Proposition 5.2 *Let A be an abelian group, $\mathbb{Z}_2 = \langle \varphi \rangle$ is the cyclic group of order 2 and φ acts on the elements of A by the following manner:*

$$a^{\varphi} = a^{-1}, \quad a \in A.$$

Then, the non-abelian tensor product $A \otimes \mathbb{Z}_2$ is defined and there is an isomorphism

$$A \otimes \mathbb{Z}_2 \cong A.$$

Proof It is not difficult to check that defined actions are compatible.

Since A acts on $\mathbb{Z}_2$ trivially and A is abelian, then the defining relations of the tensor product

$$aa_1 \otimes h = (a^{a_1} \otimes h^{a_1})(a_1 \otimes h), \quad a, a_1 \in A, \quad h \in \mathbb{Z}_2,$$

have the form

$$aa_1 \otimes h = (a \otimes h)(a_1 \otimes h) = (a_1 \otimes h)(a \otimes h). \tag{1}$$

The relations

$$a \otimes hh_1 = (a \otimes h_1)(a^{h_1} \otimes h^{h_1}), \quad a \in A, \quad h, h_1 \in \mathbb{Z}_2,$$

give only one nontrivial relation

$$1 = a \otimes \varphi^2 = (a \otimes \varphi)(a^{-1} \otimes \varphi), \quad a \in A,$$

which follows from (1).

Since the set of relations (1) is a full system of relations for $A \otimes \mathbb{Z}_2$, then there exists the natural isomorphism of $A \otimes \mathbb{Z}_2$ on A that is defined by the formula

$$a \otimes \varphi \mapsto a, \quad a \in A.$$

□

Remark 5.3 If $A = \mathbb{Z}$, then the claim of the proposition was noted in [4, Remark 4.9].

Acknowledgements The authors gratefully acknowledge the support from the RFBR-16-01-00414 and RFBR-18-01-00057. Also, we thank S. Ivanov and V. Thomas for the interesting discussions and useful suggestions.

References

1. R. Brown, J.-L. Loday, Excision homotopique en basse dimension. C. R. Acad. Sci. Paris Ser. I Math. **298**(15), 353–356 (1984)
2. R. Brown, J.-L. Loday, Van Kampen theorems for diagrams of spaces. Topology **26**(3), 311–335 (1987). with an appendix by M. Zisman
3. R. Brown, D.L. Johnson, E.F. Robertson, Some computations of non-abelian tensor products of groups. J. Algebra **111**, 177–202 (1987)
4. G. Donadze, M. Ladra, V. Thomas, More on the non-abelian tensor product and the Bogomolov multiplier. J. Algebra **472**, 399–413 (2017)
5. A.S.-T. Lue, The Ganea map for nilpotent groups. J. Lond. Math. Soc. **14**, 309–312 (1976)
6. C. Miller, The second homology group of a group; relations among commutators. Proc. AMS **3**, 588–595 (1952)
7. M.P. Visscher, On the nilpotency class and solvability length of nonabelian tensor products of groups. Arch. Math. (Basel) **73**(3), 161–171 (1999)

On Zeros of Characters of Finite Groups

Silvio Dolfi, Emanuele Pacifici and Lucia Sanus

1 Introduction

Let G be a finite group. If the character table of G is known, then some very deep structural information on G can be deduced; in fact, an important problem in character theory is to determine which structural features of G can be detected by the knowledge of the character table of G and, on the other hand, which aspects of the table are significant for this purpose.

Many results in the literature show that the *distribution of zeros* in the character table is relevant in this context. Our aim in this paper is to present an outline of this research topic. We will discuss several aspects of the subject, from classical results to recent developments, and point out some open problems that could be of interest. (For the convenience of the reader, questions and conjectures are emphasized in slanted text.)

The first and the second author are partially supported by the Italian INdAM-GNSAGA. The third author is partially supported by the Spanish MINECO proyecto MTM2016-76196-P, partly with FEDER funds, and PROMETEOII/2015/011-Generalitat Valenciana.

S. Dolfi (✉)
Dipartimento di Matematica U. Dini, Università degli Studi di Firenze,
viale Morgagni 67/a, 50134 Firenze, Italy
e-mail: dolfi@math.unifi.it

E. Pacifici
Dipartimento di Matematica F. Enriques, Università degli Studi di Milano,
via Saldini 50, 20133 Milano, Italy
e-mail: emanuele.pacifici@unimi.it

L. Sanus
Departament de Matemàtiques, Facultat de Matemàtiques, Universitat de València,
Burjassot, 46100 Valencia, Spain
e-mail: lucia.sanus@uv.es

N. S. N. Sastry and M. K. Yadav (eds.), *Group Theory and Computation*,
Indian Statistical Institute Series, https://doi.org/10.1007/978-981-13-2047-7_3

A large number of the results quoted in this paper rely on the classification of finite simple groups. This holds for virtually all the statements discussed from Sects. 4–9, except in some obvious situations (for instance, when the analysis involves only solvable groups). As for Sects. 2 and 3, we indicate explicitly the cases in which the classification comes into play.

In what follows, every group is assumed to be finite and, for the notation, we refer to [16].

2 A Theorem by W. Burnside

As one of the triggers for the research concerning zeros of characters, we recall a classical result by W. Burnside (Theorem 3.15 of [16]).

Theorem 2.1 *Let G be a group, and χ an irreducible character of G which is nonlinear (i.e., whose degree is larger than 1). Then there exists $g \in G$ such that $\chi(g) = 0$.*

This important theorem has been extended in several directions. The following result by G. Navarro yields Burnside's theorem if the subgroup N is chosen to be the trivial subgroup of G.

Theorem 2.2 ([29], Theorem A) *Let G be a group, χ an irreducible character of G, and N a normal subgroup of G. Then the restriction χ_N, which is a character of N, is not irreducible if and only if there exists $g \in G$ such that $\chi(x) = 0$ for every $x \in gN$.*

As another kind of extension for Burnside's theorem, G. Malle, G. Navarro and J.B. Olsson investigated the relationship between the "arithmetical structure" of the degree of an irreducible character and that of an element on which the character vanishes.

Theorem 2.3 ([23], Theorem B) *Let G be a group, and χ an irreducible character of G which is nonlinear. Then there exists a prime number p and a p-element $g \in G$ such that $\chi(g) = 0$.*

Now, let p be a prime number. Recalling that, whenever a character χ of the group G vanishes on a p-element of G, then the degree of χ is a multiple of p [6, Corollary 2.2], the above theorem immediately yields the following nice corollary.

Corollary 2.4 ([23], Theorem A) *Let G be a group, and χ an irreducible character of G which is nonlinear. Assume that the degree of χ is a π-number, where π is a set of primes. Then there exists a π-element $g \in G$ such that $\chi(g) = 0$.*

The two aforementioned results of [23] rely on the classification of finite simple groups.

In view of the previous statements, one may wonder whether it is true that if $\chi(1)$ is a p-power, then χ does not vanish on p'-elements. This is false in general; for instance, the Mathieu group M_{11} has an irreducible character of degree 11 which takes value 0 on an element of order 6. Moreover, a solvable example can be obtained considering the wreath product $G = C_6 \wr C_5$ of a cyclic group of order 6 with a cyclic group of order 5; it is not hard to check that G has an irreducible character of degree 5 (induced from the base group) which vanishes on an element of order 6.

Nevertheless, in the solvable context and for *primitive* characters, the situation is quite neat.

Theorem 2.5 ([28], Corollary B) *Let G be a solvable group, and χ a primitive character of G which is nonlinear. Assume that the degree of χ is a π-number, where π is a set of primes. Then, for $x \in G$, we have $\chi(x) = 0$ if and only if $\chi(x_\pi) = 0$, where x_π denotes the π-part of the element x.*

Finally, another question that may arise looking at Theorem 2.3 is the following: Does a nonlinear irreducible character always have zeros of prime order? The answer turns out to be affirmative for simple groups, as shown in [23] (in which simple groups of Lie type and sporadic simple groups are treated) and in [2] (where the authors consider alternating groups). On the other hand, the answer is negative in general: it is enough to consider the quaternion group Q_8, in which the unique element of prime order is central and therefore not a zero for any irreducible character. The next result yields some information in this context.

Theorem 2.6 ([24], Theorem A) *Let χ be a faithful irreducible character of G, and assume that $\chi(1)$ is a power of a prime p. If $\chi(x) \neq 0$ for every element $x \in G$ of order p, then the Sylow p-subgroups of G are either cyclic or generalized quaternion groups.*

Note that, if the degree of a faithful irreducible character χ is not a p-power, the condition that every element of order p is not a zero for χ does not imply that the Sylow p-subgroups of G are cyclic or generalized quaternion groups. In fact, consider $G = \mathrm{PSL}(2, 7)$; then G, which has dihedral Sylow 2-subgroups, also has an irreducible character χ of degree 6, such that $\chi(x) \neq 0$ for every involution $x \in G$.

3 Vanishing Elements

Another way of stating Theorem 2.1 is the following.

Let $\mathcal{R}$ be a row in the character table of a group G. Then $\mathcal{R}$ contains zeros if and only if $\mathcal{R}$ corresponds to a nonlinear character.

(In fact, Theorem 2.1 provides the "if" part, whereas the "only if" part is an elementary fact in character theory.) So, the problem of determining which rows in the character table of a group actually contain zeros is completely solved.

Now, if one considers the "dual" question of *which columns* in the character table of a group may contain zeros, the situation is much more complicated. In this context, the relevant objects are the so-called "vanishing elements", introduced in an important paper by I.M. Isaacs, G. Navarro and T.R. Wolf [17]: an element $g \in G$ is a vanishing element if there exists an irreducible character χ of G such that $\chi(g) = 0$. The question we are considering is therefore related to understanding which elements of a group are vanishing elements.

Given the standard duality between results concerning rows (i.e., irreducible characters) and columns (i.e., conjugacy classes) in the character table of a group, one might naively ask whether the following holds.

Let C be a column in the character table of a group G. Is it then true that C contains zeros if and only if C corresponds to a noncentral conjugacy class?

It is immediately clear that the "only if" part is true by elementary arguments, but the "if" part fails in general. In order to see it, we can just consider a 3-cycle in the symmetric group S_3: such an element is obviously noncentral and also nonvanishing in that group.

Certainly, there are special situations in which the "if" part is also true (for instance, it holds for nilpotent groups, as shown in Theorem B of [17]), but in general a nonvanishing element of G can even fail to lie in any *abelian* normal subgroup of G. Actually, in Sect. 5 of [17], the authors provide the following family of examples: for every prime p, they construct a solvable group G having nonvanishing p-elements (also, elements of order p when $p \neq 2$) which do not lie in any abelian normal subgroup of G.

However, under some suitable assumptions, a nonvanishing element of G is forced to lie in a *nilpotent* normal subgroup of G (i.e., it lies in the Fitting subgroup $\mathbf{F}(G)$). In fact, the main result of [17] is as follows.

Theorem 3.1 ([17], Theorem D) *Let G be a solvable group. If g is a nonvanishing element of G, then the image of g under the natural homomorphism onto $G/\mathbf{F}(G)$ has 2-power order.*

In particular, in a solvable group G, the nonvanishing elements *of odd order* lie in $\mathbf{F}(G)$.

In [17], *the authors actually conjecture that every nonvanishing element of a solvable group G lies in* $\mathbf{F}(G)$. They point out that their methods would prove this claim, if it can be proved that every nonvanishing element *of order* 2 of a solvable group G lies in an abelian normal subgroup of G (recall that, in [17, Sect. 5], the authors provide a counterexample to a similar statement where 2 is replaced by any odd prime). However, in a recent paper [15], M. Grüninger constructs an example of a solvable group having a nonvanishing involution which fails to lie in any abelian normal subgroup, thus showing that the prime 2 is not an exception. In any case, at the time of this writing, the conjecture by Isaacs, Navarro, and Wolf is still an open problem.

On the other hand, the assumption of solvability is certainly crucial in Theorem 3.1. If we look, for instance, at the character table of the alternating group A_7

(whose Fitting subgroup is of course trivial), we see that there are nonvanishing elements of order 2 and 6, but also of odd order (namely, of order 3).

In fact, the primes 2 and 3 do play a distinguished role in this context.

Theorem 3.2 ([11], Theorem A) *Let G be a group, and $g \in G$ an element whose order is coprime to 6. If g is a nonvanishing element of G, then $g \in \mathbf{F}(G)$.*

(The proof of the above theorem uses the classification of finite simple groups.) It seems natural to think that, for any group G, the nonvanishing elements of G should always lie in the generalized Fitting subgroup $\mathbf{F}^*(G)$, but this is not true. The group $G = 2^{11} : \mathrm{M}_{24}$ has nonvanishing elements of order 2 and 4 not lying in $\mathbf{F}(G) = \mathbf{F}^*(G)$. *However, we conjecture that any nonvanishing element of odd order of a group G lies in $\mathbf{F}^*(G)$.*

In order to give an idea of some methods that are relevant in the present context, we close this section with two easy remarks and a last theorem.

Proposition 3.3 *Let N be a normal subgroup of G, and let θ be an irreducible character of N. Then every element of G not lying in $\bigcup_{g \in G} I_G(\theta^g)$ is a vanishing element of G.*

(In the above statement, $I_G(\theta)$ denotes the inertia subgroup of θ in G, i.e., the stabilizer of θ in the natural action of G on $\mathrm{Irr}(N)$.)

Proposition 3.4 *Let N be a normal subgroup of G, and p a prime. If there exists an irreducible character of p-defect zero of N (i.e., a character $\theta \in \mathrm{Irr}(N)$ such that p does not divide $|N|/\theta(1)$), then every $g \in N$ with $p \mid o(g)$ is a vanishing element of G.*

Proposition 3.3 (whose proof is an immediate application of Clifford's Theory) is particularly useful in the case when N is an elementary abelian p-group for some prime p (for instance, when N is an abelian minimal normal subgroup of G). In this situation, the set $\mathrm{Irr}(N)$ is an elementary abelian p-group as well, and the natural action of G on this set can be regarded as a module action. By Proposition 3.3, an element $g \in G$ is a vanishing element provided, under this natural action, there exists a *deranged orbit* for g, i.e., an orbit in which no element is fixed by g. So, the study of certain orbit properties in module actions turn out to be crucial when dealing with vanishing elements.

Also, Proposition 3.4 can be proved by means of elementary character theory, taking into account that an irreducible character of p-defect zero takes value zero on every element of the group whose order is divisible by p. This proposition comes into play when N is a *nonabelian* minimal normal subgroup of G. In this case, in fact, N is a direct product $S_1 \times \cdots \times S_k$ of pairwise isomorphic nonabelian simple groups and, given a prime divisor p of $|N|$, irreducible characters of p-defect zero of N very often exist (this happens in particular whenever the S_i are simple groups of Lie type).

We note that as Proposition 3.4 may suggest, nonsolvable groups tend to have a large number of vanishing elements (for instance, by the above remarks about the existence of characters of p-defect zero, every nontrivial element of a simple group of Lie type is vanishing); in other words, a small ratio of vanishing elements in the group should imply solvability. *In fact, we conjecture that the smallest value of this ratio among nonsolvable groups are attained by the alternating group* A_7, *in which the vanishing elements are* 2134 *out of* 2520 ($\sim$85%).

The two propositions above, together with some other techniques and ideas (and the classification of finite simple groups), are used in order to prove the following theorem, which is in turn very useful for locating vanishing elements.

Theorem 3.5 ([4], Corollary 4.4). *Let A be an abelian minimal normal subgroup of G. Let N/M be a chief factor of G such that $|N/M|$ is coprime with $|A|$ and $\mathbf{C}_N(A) = M$. Then every element of $N \setminus M$ is a vanishing element of G.*

4 Ito–Michler Theorem and Vanishing Elements

An important object that can be "extracted" from the character table of a group G is the set $\mathrm{cd}(G)$, whose elements are the degrees of the irreducible characters of G. Even this relatively small set of positive integers, as shown by many results in the literature, encodes nontrivial information about the structure of G; in particular, there is a significant interplay between the group structure and the arithmetical structure of $\mathrm{cd}(G)$ (i.e., the way in which the numbers in this set decompose into prime factors). As a famous example of this relationship, we recall the celebrated Ito–Michler Theorem.

Theorem 4.1 (Ito–Michler) *Let G be a group, and p a prime. Then, every number in $\mathrm{cd}(G)$ is not divisible by p if and only if G has an abelian normal Sylow p-subgroup.*

The above statement can be regarded as a model for a certain kind of results that, following G. Navarro, we call "Ito–Michler type" theorems (see [30]). The question addressed in such theorems is the following (or a variation of it): consider a finite nonempty set X of positive integers which is attached to a group G, and assume that a given prime p does not divide any number in X; which structural properties of G can be derived as a consequence of this assumption?

Many sets of positive integers, related with a finite group G, have been considered in the literature. Among them, some classical examples are the set $\mathrm{o}(G)$ of orders of the elements of G, and the set $\mathrm{cs}(G)$ of conjugacy class sizes of G. Now, these sets can be "filtered" by means of the irreducible characters of G, in terms of the zeros appearing in the character table of G: namely, our following discussion will focus on the sets

$$\mathrm{vo}(G) = \{o(g) \mid g \text{ is a vanishing element of } G\}$$

and

$$\mathrm{vcs}(G) = \{|g^G| \mid g \text{ is a vanishing element of } G\},$$

where by g^G we denote the conjugacy class of the element g in G.

5 Ito–Michler Type Theorems: The Set vo(G)

In this section, we survey some Ito–Michler type theorems concerning the first of the two sets introduced above (or theorems that, however, relate some arithmetical properties of this set to the group structure). We start by considering the situation when, given a prime p, the set $\mathrm{vo}(G)$ does not contain any p-power.

Theorem 5.1 ([10], Theorem A) *Let G be a group, p a prime number, and P a Sylow p-subgroup of G. Assume that, for every $\chi \in \mathrm{Irr}(G)$ and $x \in P$, we have $\chi(x) \neq 0$ (i.e., assume that $\mathrm{vo}(G)$ does not contain any p-power). Then, G has a normal Sylow p-subgroup.*

The above statement (which is a consequence of Theorem 3.2 if p is larger than 3) is actually a bit stronger than a classical Ito–Michler type theorem, as the assumption that p does not divide any number in $\mathrm{vo}(G)$ clearly implies the hypothesis of Theorem 5.1.

Also the original Ito–Michler assumption that p does not divide any number in $\mathrm{cd}(G)$ implies the hypothesis of Theorem 5.1, because, as recalled in the paragraph following Theorem 2.3, an irreducible character of G which vanishes on a p-element has a degree divisible by p. On the other hand, the converse is not true. For example, let G be the normalizer of a Sylow 2-subgroup in the Suzuki group Suz(8); then G is a Frobenius group with a Frobenius complement of order 7 and a nonabelian Frobenius kernel of order 2^6. It turns out that $\mathrm{vo}(G) = \{7\}$, and $\mathrm{cd}(G) = \{1, 7, 14\}$. More generally, [6, Example 1] shows that there is no bound on the derived length of the Sylow p-subgroup of a group G such that $\mathrm{vo}(G)$ does not contain any p-power.

As an immediate consequence of Theorem 5.1, we get the following refinement of another famous result by Burnside, the so-called $p^\alpha q^\beta$ Theorem.

Theorem 5.2 ([10], Corollary B) *Let G be a group, and let p, q be prime numbers. If every vanishing element of G is a $\{p, q\}$-element, then G is solvable.*

In the next result, the hypothesis of Theorem 5.1 is relaxed, assuming only that $\mathrm{vo}(G)$ does not contain the prime p.

Theorem 5.3 ([10], Theorem 4.3) *Let G be a group, and p a prime divisor of $|G|$. Assume that either p is odd, or that $p = 2$ and G has no composition factor isomorphic to M_{22}, A_7 or A_{15}. If $\mathrm{vo}(G)$ does not contain p, then $\mathbf{O}_p(G) \neq 1$.*

Before we proceed in our discussion related to Ito–Michler type theorems, we take some time to consider the opposite situation in which the set $\mathrm{vo}(G)$ only contains p-powers, or it even reduces to a single prime number.

Theorem 5.4 ([6], Theorem A) *Let G be a nonabelian group, and p a prime. If every number in* $\mathrm{vo}(G)$ *is a p-power, then one of the following holds.*

(a) *G is a p-group.*
(b) *$G/\mathbf{Z}(G)$ is a Frobenius group with a Frobenius complement of p-power order and $\mathbf{Z}(G) = \mathbf{O}_p(G)$.*

Theorem 5.5 ([6], Theorem B) *Let G be a nonabelian group, and p a prime. If* $\mathrm{vo}(G) = \{p\}$*, then one of the following holds.*

(a) *G is a p-group of exponent p.*
(b) *$G = E \times F$, where E is a (possibly trivial) elementary abelian p-group and F is a Frobenius group with a Frobenius complement of order p.*

Theorem 5.6 ([6], Theorem C) *Let G be a nonabelian group. Then,* $\mathrm{vo}(G) = \{2\}$ *if and only if $G = E \times F$, where E is an elementary abelian* 2*-group and F is a Frobenius group with a Frobenius complement of order* 2.

Finally, we resume the discussion about Theorem 5.1 by observing that, as shown by any nonabelian p-group, the converse of that statement is false. In other words, Theorem 5.1 does not provide a characterization of normality for a Sylow p-subgroup in terms of the character table of G.

The problem of achieving such a characterization along this line was considered by G. Malle and G. Navarro in [22]. In that paper, the authors introduce one particular set of irreducible characters of a group: given a group G, a prime p and a Sylow p-subgroup P of G, they define

$$\mathrm{Irr}((1_P)^G) = \{\chi \in \mathrm{Irr}(G) \mid \langle \chi_P, 1_P \rangle \neq 0\},$$

i.e., the subset of $\mathrm{Irr}(G)$ whose elements are the irreducible constituent of the character of G obtained by inducing the principal character of P.

Our discussion concerning Theorem 5.1 yields

$$p \nmid \chi(1) \text{ for every } \chi \in \mathrm{Irr}(G) \Rightarrow \chi(x) \neq 0 \text{ for every } \chi \in \mathrm{Irr}(G) \text{ and } x \in P \Rightarrow P \trianglelefteq G.$$

Now, if every occurrence of $\mathrm{Irr}(G)$ in the previous line is replaced by $\mathrm{Irr}((1_P)^G)$, then both the implications are in fact “if and only if”.

Theorem 5.7 ([22], Theorem B) *Let G be a group, p a prime number, and P a Sylow p-subgroup of G. Then, the following conditions are equivalent.*

(a) *For every χ in* $\mathrm{Irr}((1_P)^G)$*, the prime p does not divide $\chi(1)$.*
(b) *For every χ in* $\mathrm{Irr}((1_P)^G)$ *and $x \in P$, we have $\chi(x) \neq 0$.*
(c) *$P \trianglelefteq G$.*

Therefore, while Ito–Michler Theorem yields a characterization of *normality and abelianity* of a Sylow p-subgroup in terms of the character table, the theorem above provides a neat characterization of *normality* for a Sylow p-subgroup in terms of the character table (namely, in terms of degrees and of the distribution of zeros in the character table).

6 Ito–Michler Type Theorems: The Set vcs(G)

In the same spirit as in the previous section, we now focus on the set of conjugacy class sizes of a group. First of all, we state the classical Ito–Michler type theorem on the whole set of class sizes, whose proof is an elementary exercise.

Theorem 6.1 *Let G be a group, and p a prime number. Then p does not divide any number in* $\mathrm{cs}(G)$ *if and only if G has a central Sylow p-subgroup (i.e., G has a p-complement H that is a direct factor, and G/H is abelian).*

What if the Ito–Michler assumption is required only for the sizes of the vanishing conjugacy classes? In this case, the right idea is to focus on the *principal p-block* (see for instance [27, p. 49]). In fact, the following lemma turns out to be a crucial one.

Lemma 6.2 *Let G be a group, p a prime, and B_0 the principal p-block of G. If χ is an (ordinary) irreducible character of G lying in B_0, and $x \in G$ is such that $\chi(x) = 0$, then p divides $|x^G|$.*

Proof Let $\mathbf{R}$ be the ring of algebraic integers in the complex field, and let M be a fixed maximal ideal of $\mathbf{R}$ containing the ideal generated by p. Also, denote by * the natural homomorphism of $\mathbf{R}$ onto the field $\mathbf{R}/M$.

By definition, since the irreducible character χ of G lies in B_0, we get

$$\left(\frac{|g^G| \cdot \chi(g)}{\chi(1)}\right)^* = |g^G|^*$$

for every $g \in G$. In particular, as $\chi(x) = 0$, we have that $|x^G|$ is an integer lying in $p\mathbf{R}$; it follows that p divides $|x^G|$, as claimed. ■

Define now

$$\mathrm{Van}(B_0) = \{x \in G \mid \chi(x) = 0 \text{ for some } \chi \in \mathrm{Irr}(B_0)\}.$$

As an immediate consequence of Lemma 6.2, we obtain the following result.

Theorem 6.3 *Let G be a group, and p a prime number. Then, p does not divide $|x^G|$ for every $x \in$ $\mathrm{Van}(B_0)$ if and only if G has a normal p-complement H and G/H is abelian.*

Proof If G has a normal p-complement H, then [27, Theorem 6.10] yields that H is the intersection of the kernels of all the irreducible ordinary characters in B_0. Therefore, if G/H is assumed to be abelian, these characters are in fact linear, so that $\mathrm{Van}(B_0)$ is empty and nothing else needs to be proved.

Conversely, if p does not divide $|x^G|$ for every $x \in \mathrm{Van}(B_0)$, then Lemma 6.2 (together with Burnside's Theorem 2.1) yields that the irreducible characters in B_0 are all linear. Now, again Theorem 6.10 of [27] implies that G' lies in $\mathbf{O}_{p'}(G)$ (the maximal normal p'-subgroup of G), and the desired conclusion easily follows. ■

Thus, *if G is a group and p is a prime which does not divide any number in* $\mathrm{vcs}(G)$, *then G has a normal p-complement H and G/H is abelian.* In fact, in [5, Theorem C], the author obtains a strengthening of this statement: if p does not divide the class size *of any vanishing p'-element of G*, then G has a normal p-complement with abelian factor group.

Observe that the structural information which is lost in this context, with respect to the stronger assumptions of Theorem 6.1, concerns *the normality* of a Sylow p-subgroup of G. In fact, in the symmetric group $G = \mathrm{S}_3$, the class of transpositions is the unique vanishing conjugacy class. This class has size 3, therefore the prime 2 does not divide any number in $\mathrm{vcs}(G)$; nevertheless, G does not have a normal Sylow 2-subgroup.

Assume now that, for a given prime p, the group G has a p-complement H, and let us define

$$\mathrm{Van}(G \mid 1_H) = \{x \in G \mid \chi(x) = 0 \text{ for some } \chi \in \mathrm{Irr}(G) \text{ with } \langle \chi_H, 1_H \rangle \neq 0\}.$$

Taking into account that every irreducible constituent of the induced character $(1_H)^G$ lies in B_0 (see [27, Theorem 2.27]) and arguing along the line of Theorem 6.3, it is not difficult to prove the following result, that is very much in the spirit of the work by Malle and Navarro in [22].

Theorem 6.4 *Let p be a prime, and G a group having a p-complement H. Then p does not divide $|x^G|$ for every $x \in \mathrm{Van}(G \mid 1_H)$ if and only if $H \trianglelefteq G$ and G/H is abelian.*

We close this section remarking that, again in the spirit of the work by Malle and Navarro in [22], *it could be interesting to find a characterization of p-nilpotency for a group G (i.e., the existence of a normal p-complement $H \leq G$, but without any extra condition on G/H) in terms of vanishing conjugacy classes.*

7 Vanishing Graphs

Given a nonempty finite set X of positive integers, a way to express the arithmetical properties of the integers in X is as follows. Consider the so-called *prime graph* on X, that is the simple undirected graph $\Delta(X)$ with vertex set

$$\mathrm{V}(\Delta(X)) = \{p \text{ prime} \mid \text{there exists } x \in X \text{ divisible by } p\},$$

and define two vertices p, q to be adjacent in $\Delta(X)$ if there exists an integer $x \in X$ such that pq divides x.

(Similarly, another graph that comes naturally into consideration is the "common divisor graph" $\Gamma(X)$, whose vertex set is $X \setminus \{1\}$ and $x, y \in X \setminus \{1\}$ are connected if $\gcd(x, y) \neq 1$.)

The main general question in this context is how the group structure of G is related to the structure of the corresponding graphs $\Delta(X)$, for various sets X of invariants of a group G.

As mentioned in Sect. 3, one of the earliest instances is the set $X = \mathrm{o}(G)$ consisting of the orders of the elements of the group G. The corresponding graph $\Pi(G) = \Delta(\mathrm{o}(G))$ is called the *Gruenberg–Kegel graph* and it has been extensively studied both in the solvable as well as in the nonsolvable case.

Among the various graph properties, the most commonly studied in the present literature is related to the diameter and the number of connected components. In the following discussion, given a graph Δ, we denote by $\mathrm{n}(\Delta)$ the number of connected components of Δ and by $\mathrm{diam}(\Delta)$ its diameter. Finally, we denote by $\iota(\Delta)$ the independence number of Δ, that is the largest size of an independent set, i.e., a subset of pairwise nonadjacent vertices of Δ.

We recall that a group G is said to be a 2-Frobenius group if there exist two normal subgroups F and L of G such that L is a Frobenius group with kernel F, and G/F is a Frobenius group with kernel L/F. For the Gruenberg–Kegel graph of solvable groups, we have:

Theorem 7.1 ([20, 34]) *Let G be a solvable group.*

(a) $\mathrm{n}(\Pi(G)) \leq 2$, *i.e.,* $\Pi(G)$ *has at most two connected components.*
(b) *If* $\Pi(G)$ *is disconnected, then G is either a Frobenius or a 2-Frobenius group and each connected component of* $\Pi(G)$ *is a complete graph.*
(c) *For any choice of three vertices of* $\Pi(G)$, *at least two of them are adjacent in* $\Pi(G)$ *(i.e.,* $\iota(\Pi(G)) \leq 2$*).*

Aiming at filtering the elements of the set $\mathrm{o}(G)$ by properties related to character values, in Sect. 4 we introduced the set $\mathrm{vo}(G)$ consisting of the orders of the vanishing elements of G. Accordingly, one defines the *vanishing* Gruenberg–Kegel graph $\Pi_v(G) = \Delta(\mathrm{vo}(G))$ of G as the prime graph on the set of the orders of the vanishing elements of G. Clearly, $\Pi_v(G)$ is a subgraph of $\Pi(G)$. Still, it is not an *induced* subgraph: as an example, consider $G = \mathrm{S}_3 \times D_{10}$, where 3 and 5 are vertices of $\Pi_v(G)$ which are linked in $\Pi(G)$, but not in $\Pi_v(G)$.

In the process of comparing $\Pi(G)$ and $\Pi_v(G)$, one can first ask about the difference between the vertex sets $\mathrm{V}(\Pi(G))$ and $\mathrm{V}(\Pi_v(G))$.

Theorem 7.2 ([12]) *Let G be a nonabelian group, p a prime number, and* $P \in \mathrm{Syl}_p(G)$. *If p is a vertex of* $\Pi(G)$ *but not of* $\Pi_v(G)$, *then* $P \trianglelefteq G$, $G/\mathbf{O}_{p'}(G)$ *is a Frobenius group with kernel* $P\mathbf{O}_{p'}(G)/\mathbf{O}_{p'}(G)$ *and* $\mathbf{O}_{p'}(G)$ *is nilpotent.*

We say that a group G is a *nearly* 2-*Frobenius group* if there exist two normal subgroups F and L of G with the following properties: $F = F_1 \times F_2$ is nilpotent, where F_1 and F_2 are normal subgroups of G, G/F is a Frobenius group with kernel L/F, G/F_1 is a Frobenius group with kernel L/F_1, and G/F_2 is a 2-Frobenius group. The next result should be compared with Theorem 7.1.

Theorem 7.3 ([13]) *Let G be a solvable group. Then the following conclusions hold.*

(a) *$\Pi_v(G)$ has at most two connected components. If $\Pi_v(G)$ is disconnected, then each component is a complete graph, and G is a Frobenius or a nearly 2-Frobenius group.*
(b) $\mathrm{diam}(\Pi_v(G)) \leq 4$.

We remark that the bound $\mathrm{diam}(\Pi_v(G)) \leq 4$ is sharp [13, Example 5.2].

By contrast, the similarity of the ordinary and vanishing Gruenberg–Kegel graphs breaks down when one considers independence numbers: while one has independent sets of maximal size two, the other can have arbitrarily large independent sets.

Theorem 7.4 ([13], Theorem B) *For every positive integer k, there exists a solvable group G such that $\Pi_v(G)$ has an independent set of size k.*

Removing the assumption of solvability, from [18, 34] it is possible do derive the following result.

Theorem 7.5 (a) *If S is a nonabelian simple group, then* $\mathrm{n}(\Pi(S)) \leq 6$.
(b) *Let G be a nonsolvable group. If $\Pi(G)$ is disconnected, then G has a unique nonabelian composition factor S, and* $\mathrm{n}(\Pi(G)) \leq \mathrm{n}(\Pi(S))$. *Hence,* $\mathrm{n}(\Pi(G)) \leq 6$.

Similarly, for the vanishing Gruenberg–Kegel graph:

Theorem 7.6 ([12], Theorem A) *Let G be a finite group. Then, the following conclusions hold.*

(a) *$\Pi_v(G)$ has at most six connected components.*
(b) *If $\Pi_v(G)$ is disconnected, then G has a unique nonabelian composition factor S, and* $\mathrm{n}(\Pi_v(G)) \leq \mathrm{n}(\Pi(S))$ *unless G is isomorphic to A_7.*

In fact, it turns out that A_7 is the unique nonabelian simple group S such that $\Pi_v(S) \neq \Pi(S)$. Note that $\mathrm{n}(\Pi_v(\mathrm{A}_7)) = 4$, while $\mathrm{n}(\Pi(\mathrm{A}_7)) = 3$.

We stress that notwithstanding the similarities among the two graphs, the edge set in the graph $\Pi_v(G)$ can be quite smaller than in the graph $\Pi(G)$; for any integer k, there exists a (nonsolvable) group G such that $\Pi(G)$ has a complete subgraph on k vertices, that instead induces an independent set in $\Pi_v(G)$ [12, Example 6.5].

Other graph properties, like connectivity, chromatic number, or girth, might be subjects for further investigation.

Also the arithmetical properties of the sets $\mathrm{cd}(G)$ and $\mathrm{cs}(G)$, that have been introduced in Sect. 3, can be studied via the prime graph. Several properties of the graphs $\Delta(\mathrm{cd}(G))$ and $\Delta(\mathrm{cs}(G))$, as well as their connection to the algebraic structure of the group, have been studied in the last two decades. For an overview up to 2008, we refer to the survey paper [19].

In the same spirit, we now focus on the set $\mathrm{vcs}(G)$ of the sizes of vanishing classes. As we observed in Sect. 6, if a prime number p is not a vertex of $\Delta(\mathrm{vcs}(G))$, then G has a normal p-complement and abelian Sylow p-subgroups. We also observed that the vertex set of $\Delta(\mathrm{vcs}(G))$ can be smaller than that of $\Delta(\mathrm{cs}(G))$. Yet, if one assumes that G has a nonabelian minimal normal subgroup, then the two vertex sets coincide, as proved in [4]. In this situation, the absence of an edge in the graph $\Delta(\mathrm{vcs}(G))$ reflects in the normal structure of G.

Theorem 7.7 ([4], Theorem A) *Let G be a finite group, and suppose that G has a nonabelian minimal normal subgroup. If p and q are vertices of $\Delta(\mathrm{vcs}(G))$, but there is no vanishing conjugacy class of G whose size is divisible by pq, then G is $\{p, q\}$-solvable.*

We remark that the assumption concerning the existence of a nonabelian minimal normal subgroup in G is critical in the above statement. In fact, whenever p and q are primes such that $p \geq 7$ and $q \equiv 1 \pmod{5p}$, it is possible to construct a Frobenius group H whose kernel is elementary abelian of order q^2 and whose complements are isomorphic to $C_p \times \mathrm{SL}(2, 5)$; it is not difficult to see that p is not a vertex in $\Delta(\mathrm{vcs}(H))$. Now, take $p = 7, q = 71$, and consider $G = \mathrm{D}_{10} \times H$ (where D_{10} is the dihedral group of order 10); clearly, 2 and 7 are nonadjacent vertices in $\Delta(\mathrm{vcs}(G))$ (although they are adjacent in $\Delta(\mathrm{cs}(G))$), nevertheless G is not 2-solvable. (The authors whish to thank Victor Manuel Ortiz Sotomayor for pointing out this kind of examples; a solvable one is $G = \mathrm{D}_{10} \times \mathrm{A}_4$, in which 2 and 3 are nonadjacent vertices of $\Delta(\mathrm{vcs}(G))$ that are adjacent in $\Delta(\mathrm{cs}(G))$.)

A consequence of the previous theorem is that still assuming the existence of a nonabelian minimal normal subgroup in G, if a vertex p of $\Delta(\mathrm{vcs}(G))$ is not complete (i.e., adjacent to all other vertices), then the group G is p-solvable.

Moreover, if the group has no abelian normal subgroup, then the graph $\Delta(\mathrm{vcs}(G))$ is complete.

Theorem 7.8 ([4], Theorem B) *Let G be a finite group with trivial Fitting subgroup. Then, every prime divisor of $|G|$ is a vertex of $\Delta(\mathrm{vcs}(G))$, and $\Delta(\mathrm{vcs}(G))$ is a complete graph.*

We are not aware of any examples where, under the assumption that G has a nonabelian minimal normal subgroup, two primes p and q are vertices of $\Delta(\mathrm{vcs}(G))$ that are not adjacent in this graph, but adjacent in $\Delta(\mathrm{cs}(G))$. In other words, it is an open question whether in this case $\Delta(\mathrm{vcs}(G)) = \Delta(\mathrm{cs}(G))$.

8 The Number of Conjugacy Classes of Vanishing Elements

Given an irreducible character χ of G, we define $v(\chi) = |\{x^G : \chi(x) = 0\}|$, the number of zero entries in the row corresponding to χ in the character table of G. By Burnside's theorem, $v(\chi) = 0$ if and only if χ is a linear character.

It is natural to ask how the largest number

$$M(G) = \max_{\chi \in \mathrm{Irr}(G)} v(\chi)$$

of zeros in a row of the character table of G is related to the structure of G.

Theorem 8.1 ([25], Theorem A) *There exist two real numbers c_1 and c_2 such that, for every solvable group G with $M(G) > 1$,*

$$h(G) \leq c_1 \log\log M(G) + c_2 ,$$

where $h(G)$ is the Fitting height.

In [25], Moretó and Sangroniz also prove that the index of suitable terms of the Fitting series of a solvable group G can be bounded in terms of $M(G)$ [25, Theorem B]. Furthermore, the order of a nilpotent group can be bounded by some function of $M(G)$. This is not true in general, as the dihedral groups show.

Similarly, one can consider the minimum number of zeros

$$m(G) = \min_{\chi \in \mathrm{Irr}(G), \chi(1)>1} v(\chi)$$

appearing in the rows of the character table of a group G. Moretó and Sangroniz prove that the derived length of a p-group P can be bounded by a function of $m(P)$ [25, Theorem E]. They also propose the following conjectures.

Conjecture 8.2 ([25], Conjectures F and G) Let G be a solvable group. Then

(a) the derived length $dl(G)$ and the index $|G : \mathbf{F}(G)|$ can be bounded in terms of $M(G)$;
(b) the Fitting height $h(G)$ can be bounded in terms of $m(G)$.

The finite groups whose irreducible characters vanish on "few" conjugacy classes have been classified.

Theorem 8.3 ([1], Theorem 5; [7], Proposition 2.7) $M(G) = 1$ *if and only if G is a Frobenius group with complement of order* 2.

Theorem 8.4 ([3], Theorem 1.1; [25], Theorem H) $M(G) = 2$ *if and only if* $G \simeq \mathrm{S}_4, \mathrm{A}_5, \mathrm{PSL}(2, 7)$, *or there is a normal subgroup N with $M(G/N) = 1$ and $|N| = 2$ or G is a Frobenius group with complement of order* 3 *and abelian kernel.*

Finally, a classification of the groups G such that $M(G) = 3$ is given in [31].

Dually, looking at the columns of the character table of a group G, one defines $v^*(g) = |\{\chi \in \mathrm{Irr}(G) \;:\; \chi(g) = 0\}|$ and $M^*(G) = \max_{g \in G}\{v^*(g)\}$.

Theorem 8.5 ([26, Theorem A]) *The number of nonlinear irreducible characters of G is bounded in terms of $M^*(G)$. Hence, if G is solvable, the derived length $dl(G)$ is bounded above by $M^*(G)$.*

In [33], one finds a complete classification of the groups G such that $M^*(G) < p$, where p is the smallest prime divisor of $|G|$; they are either isomorphic to A_5 or they belong to one of seven families of solvable groups [33, Theorem 1.1].

A natural question in this context is about groups that have "few" orbits of vanishing conjugacy classes, or of conjugacy classes that are zeros for single irreducible characters, under some natural actions (e.g., Galois conjugation).

Finally, we mention a result that outlines a connection between rows and columns (from the point of view of zero entries) in a character table.

Theorem 8.6 ([32]) *For any finite group G, the following conditions are equivalent.*

(a) $v(\chi) \leq 1$ *for all but one of the irreducible characters* χ *of* G*;*
(b) $v^*(x^G) \leq 1$ *for all but one of the conjugacy classes of* G.

Moreover, G satisfies one of the above condiditions if and only if G is one of the following groups:

- *an extra special* 2*-group;*
- $\mathrm{SL}(2, 3)$, S_4 *or* A_8*;*
- *a Frobenius group which is either* 2*-transitive with an abelian complement or it has a complement of order* 2.

9 Brauer Characters

Unlike ordinary characters, it is possible that a nonlinear irreducible Brauer character does not vanish on any element. For instance, in characteristic 7, the irreducible Brauer characters of $\mathrm{PSL}(3, 2)$ of degree 5 and 7 do not take the value 0.

Even more, there exist nonabelian groups G whose Brauer character table, in some characteristic p, does not contain any zeros: consider for instance $G = \mathrm{S}_4$ and $p = 3$. However, for odd characteristic, this phenomenon can only happen when G is a solvable group (see Theorem 9.2 below).

The next result shows that, for p odd, all nonabelian simple groups have an irreducible p-Brauer character that vanishes on a full $\mathrm{Aut}(G)$-orbit of p-regular elements.

Theorem 9.1 ([21, Theorem 1.1]) *Let G be a nonabelian simple group and p a prime. Then, there exists a $\phi \in \mathrm{IBr}_p(G)$ and a p-regular $g \in G$ such that $\phi(g^\alpha) = 0$ for all $\alpha \in \mathrm{Aut}(G)$, unless $p = 2$ and*

- $G = L_2(2^m)$, $m \geq 2$;
- $G = L_2(q)$, $q = 2^m + 1$, $m \geq 2$;
- $G = {}^2B_2(2^{2m+1})$, $m \geq 1$;
- $G = S_4(2^m)$, $m \geq 2$.

In these cases, the degrees of all irreducible 2*-Brauer characters of G are powers of* 2.

From the above result, one derives the following:

Theorem 9.2 ([21, Theorem 1.3]) *Assume that G is not solvable and $p \neq 2$. Then there exists an irreducible p-Brauer character of G which vanishes on some p-regular element of G.*

An open question in this context is whether the degrees of the irreducible Brauer characters of a group G are necessarily all 2-powers if the 2-Brauer character table of G has no zeros.

It is natural to guess that solvable groups whose p-Brauer character table has no zeros, must have a structure of somewhat restricted type relatively to the prime p. There are examples of such groups with both p-length ($l_p(G)$) and p'-length ($l_{p'}(G)$) equal to 2 [8, Example 4.1], but this is (for $p \neq 3$) the worst it can get.

Theorem 9.3 ([8, 9]) *Let p be prime and let G be a finite group such that the p-Brauer character table of G contains no zeros. Then,*

(a) *If $p \geq 5$, then the Hall p'-subgroups of the factor group $G/\mathbf{F}(G)$ are abelian; so, $l_{p'}(G/\mathbf{F}(G)) \leq 1$, $l_{p'}(G) \leq 2$ and $l_p(G/\mathbf{O}_p(G)) \leq 2$.*
(b) *If $p = 3$, then $G/\mathbf{F}(G)$ is a subgroup of a direct product $A \times B$, where A is a $\{2,3\}$-group with elementary abelian Sylow 2-subgroups and $3'$-length at most 1 and $B \simeq (\mathrm{Sym}(3) \wr \mathrm{Sym}(3)) \wr P$, where P is a 3-group. In particular, $l_{3'}(G) \leq 3$, $l_3(G/\mathbf{O}_3(G)) \leq 3$.*
(c) *If $p = 2$ and G is solvable, then $G/\mathbf{F}(G)$ is a $\{2,3\}$-group with elementary abelian Sylow 3-subgroups; also, $l_{2'}(G) \leq 2$, $l_2(G/\mathbf{O}_2(G)) \leq 2$.*
(d) *If $p = 2$ and G is nonsolvable, then there exist normal subgroups R, N of G, $R \leq N$, with R solvable, $l_{2'}(R) \leq 4$, N/R a direct product of simple groups as listed in Theorem 9.1 and G/N a group of 2-power order.*

We remark that no examples are known of groups with no zeros in the 3-Brauer character table and with $3'$-length greater than 2. So, part (b) of the above theorem can possibly be improved.

Let p be a prime; a p-regular element of a group G is called a p-nonvanishing element of G if no irreducible p-Brauer character of G takes value zero on it. The following statement, which strengthens Theorem 9.3 for $p > 7$, locates p-nonvanishing elements of a solvable group G with respect to the p-series of G. It should be compared with Theorem 3.1.

Theorem 9.4 ([14, Theorem A]) *Let p be a prime number greater than 3, let G be a finite solvable group with $\mathbf{O}_p(G) = 1$, and let g be a p-regular element of G that is p-nonvanishing. Then, g lies in $\mathbf{O}_{p'pp'}(G)$, unless $p \in \{5, 7\}$ and the order of g is divisible by 2 or 3.*

It is unknown, at the moment, whether the assumption $p > 7$ is really needed in the above statement. The ideas used in [14] break down for small primes, but other methods could take over. *Another issue that is wide open concerns the distribution of p-nonvanishing elements in nonsolvable groups.*

References

1. Y. Berkovich, L. Kazarin, Finite groups in which the zeros of every nonlinear irreducible character are conjugate modulo its kernel. Houston J. Math. **24**, 619–630 (1998)
2. C. Bessenrodt, J.B. Olsson, Weights of partitions and character zeros. Electron. J. Combin. **11** (2004). Research Paper 5
3. M. Bianchi, D. Chillag, A. Gillio, Finite groups in which every irreducible character vanishes on at most two conjugacy classes. Houston J. Math. **26**, 451–461 (2000)
4. M. Bianchi, J.M.A. Brough, R.D. Camina, E. Pacifici, On vanishing class sizes in finite groups. J. Algebra **489**, 446–459 (2017)
5. J.M.A. Brough, On vanishing criteria that control finite group structure. J. Algebra **458**, 207–215 (2016)
6. D. Bubboloni, S. Dolfi, P. Spiga, Finite groups whose irreducible characters vanish only on p-elements. J. Pure Appl. Algebra **213**, 370–376 (2009)
7. D. Chillag, On zeros of characters of finite groups. Proc. Am. Math. Soc. **127**, 977–983 (1999)
8. S. Dolfi, E. Pacifici, Zeros of Brauer characters and linear actions of finite groups. J. Algebra **340**, 104–113 (2011)
9. S. Dolfi, E. Pacifici, Zeros of Brauer characters and linear actions of finite groups: small primes. J. Algebra **399**, 343–357 (2014)
10. S. Dolfi, E. Pacifici, L. Sanus, P. Spiga, On the orders of zeros of irreducible characters. J. Algebra **321**, 345–352 (2009)
11. S. Dolfi, G. Navarro, E. Pacifici, L. Sanus, P.H. Tiep, Non-vanishing elements of finite groups. J. Algebra **323**, 540–545 (2010)
12. S. Dolfi, E. Pacifici, L. Sanus, P. Spiga, On the vanishing prime graph of finite groups. J. London Math. Soc. **82**, 167–183 (2010)
13. S. Dolfi, E. Pacifici, L. Sanus, P. Spiga, On the vanishing prime graph of solvable groups. J. Group Theory **13**, 189–206 (2010)
14. S. Dolfi, E. Pacifici, L. Sanus, Nonvanishing elements for Brauer characters. J. Aust. Math. Soc. **102**, 96–107 (2017)
15. M. Grüninger, Two remarks about non-vanishing elements in finite groups. J. Algebra **460**, 366–369 (2016)
16. I.M. Isaacs, *Character Theory of Finite Groups*, AMS Chelsea Publishing Series (2006)
17. I.M. Isaacs, G. Navarro, T.R. Wolf, Finite group elements where no irreducible character vanishes. J. Algebra **222**, 413–423 (1999)
18. A.S. Kondrat'ev, On prime graph components of finite simple groups. Math. USSR-Sb **67**, 235–247 (1990)
19. M. Lewis, An overview of graphs associated with character degrees and conjugacy class sizes in finite groups. Rocky Mountain J. Math. **38**, 175–211 (2008)
20. M.S. Lucido, The diameter of the prime graph of a finite group. J. Group Theory **2**, 157–172 (1999)
21. G. Malle, Zeros of Brauer characters and the defect zero graph. J. Group Theory **13**, 171–187 (2010)
22. G. Malle, G. Navarro, Characterizing normal Sylow p-subgroups by character degrees. J. Algebra **370**, 402–406 (2012)
23. G. Malle, G. Navarro, J.B. Olsson, Zeros of characters of finite groups. J. Group Theory **3**, 353–368 (2000)
24. A. Moretó, G. Navarro, Zeros of characters on prime order elements. Commun. Algebra **29**, 5171–5173 (2001)
25. A. Moretó, J. Sangroniz, On the number of conjugacy classes of zeros of characters. Israel J. Math. **142**, 163–187 (2004)
26. A. Moretó, J. Sangroniz, On the number of zeros in the columns of the character table of a group. J. Algebra **279**, 726–736 (2004)
27. G. Navarro, *Characters and Blocks of Finite Groups*, LMS Lecture Note Series 250 (Cambridge University Press, Cambridge, 1998)

28. G. Navarro, Zeros of primitive characters in solvable groups. J. Algebra **221**, 644–650 (1999)
29. G. Navarro, Irreducible restrictions and zeros of characters. Proc. Amer. Math. Soc. **129**, 1643–1645 (2001)
30. G. Navarro, Variations on the Ito-Michler theorem on character degrees. Rocky Mountain J. Math. **46**, 1363–1377 (2016)
31. J. Shi, Z. Shen, J. Zhang, Finite groups whose irreducible characters vanish on at most three conjugacy classes. J. Group Theory **13**, 799–819 (2010)
32. W. Shi, J. Zhang, Two dual questions on zeros of characters of finite groups. J. Group Theory **11**, 697–708 (2008)
33. Z. Shen, S. Wu, J. Zhang, A note on the number of zeros in the columns of the character table of a group. J. Algebra Appl. **12**, 1250150 [1–10] (2013)
34. J.S. Williams, Prime graph components of finite groups. J. Algebra **69**, 487–513 (1981)

Properties of Finite and Periodic Groups Determined by Their Element Orders (A Survey)

Marcel Herzog, Patrizia Longobardi and Mercede Maj

1 Introduction

Let G be a finite group. The function

$$\omega(G) = \{o(x) : x \in G\}$$

assigns to G the set of orders of all elements of G. In this survey, we shall describe results which answer the following question:

Question 1.1 *What information about G can be derived by looking either at $\omega(G)$ or at the complete list of orders of elements of G?*

We shall accompany these results by numerous relevant open questions and conjectures. The information to which we shall refer includes the set $\omega(G)$, the subsets $L_e(G) := \{x \in G : x^e = 1\}$ of roots of unity in G for all divisors e of $|G|$, sum of the orders of all elements of G denoted by $\psi(G)$, product of these orders denoted by $P(G)$, and other functions of the orders of all elements of G. With respect to $\omega(G)$, we shall also consider periodic groups. Recall that a group is called periodic if all of its elements are of finite order.

This work was supported by the "National Group for Algebraic and Geometric Structures, and their Applications" (GNSAGA - INDAM), Italy.
The first author is grateful to the Department of Mathematics of the University of Salerno for its hospitality and support, while this investigation was carried out.

M. Herzog
School of Mathematical Sciences, Tel-Aviv University, Ramat-Aviv,
Tel-Aviv, Israel

P. Longobardi (✉) · M. Maj
Dipartimento di Matematica, Università di Salerno, via Giovanni Paolo II, 132,
84084 Fisciano (Salerno), Italy
e-mail: plongobardi@unisa.it

N. S. N. Sastry and M. K. Yadav (eds.), *Group Theory and Computation*,
Indian Statistical Institute Series, https://doi.org/10.1007/978-981-13-2047-7_4

In this survey, G, n, C_n, and p will denote a group, a positive integer, a cyclic group of order n, and a prime number, respectively. S_n and A_n will denote the symmetric and the alternating group on n letters, while Q_8 will be the quaternion group of order 8 and D_n the dihedral group of order n. We will refer to [56] for other notation. The order of $x \in G$ will be denoted by $o(x)$.

This survey consists of the following sections:

1. Introduction
2. Properties of periodic groups G determined by $\omega(G)$
3. Roots of unity in finite groups
4. Results concerning the function $\psi(G)$
5. Results concerning some other functions.

We conclude this introduction by listing some results, conjectures, and questions of special interest, which will be discussed in this survey. In the first four entries of the list, G denotes a **periodic group**.

Theorem 2.1.(2) If $\omega(G) = \{1, 2, 3\}$, then G is (elementary abelian)-by-(prime order).

Theorem 2.1.(3) If $\omega(G) \subseteq \{1, 2, 3, 4\}$, then G is locally finite.

Question 2.2.(1) Does $\omega(G) = \{1, 5\}$ imply that G is locally finite?

Question 2.2.(2) Does $\omega(G)$ being finite imply G being locally finite? This is the famous Burnside problem, which was answered negatively by Novikov and Adjan in 1968 (see [54]).

In the following entries of the list, G will denote a **finite group**. Recall that

$$L_e(G) := \{x \in G : x^e = 1\} \text{ for } e \text{ dividing } |G|,$$

$$\psi(G) := \sum_{x \in G} o(x)$$

and

$$P(G) := \prod_{x \in G} o(x).$$

We shall also consider the functions

$$R_G(r, s) := \sum_{x \in G} \frac{o(x)^s}{\varphi(o(x))^r},$$

where r, s are real numbers and φ denotes the Euler's totient function and

$$R_G(r) := \sum_{x \in G} \left(\frac{o(x)}{\varphi(o(x))}\right)^r.$$

In particular $R_G(0, 1) = \sum_{x\in G} o(x) = \psi(G)$ and

$$R_G(1) = \sum_{x\in G} \frac{o(x)}{\varphi(o(x))}.$$

Question 3.2.4 Let G be a soluble group. Is it true that if H is a finite group satisfying $|L_e(H)| = |L_e(G)|$ for every $e \in \mathbb{N}$, then H is soluble?

Theorem 4.2.1 If $|G| = n$ then $\psi(G) \leq \psi(C_n)$, with equality if and only if $G \cong C_n$.

Theorem 4.3.1 If G is noncyclic and $|G| = n$, then $\psi(G) \leq \frac{7}{11}\psi(C_n)$.

Theorem 4.3.3 If G is noncyclic, $|G| = n$ and q is the smallest prime divisor of n, then $\psi(G) < \frac{1}{q-1}\psi(C_n)$.

Corollary 4.3.4 If G is noncyclic group of odd order n, then $\psi(G) < \frac{1}{2}\psi(C_n)$.

Theorem 4.3.6 Let n be an integer larger than 1, with the largest prime divisor p and the smallest prime divisor q. Then $\varphi(n) \geq \frac{q-1}{p}n$.

Conjecture 4.6.5 If G is a soluble group and S is a simple group satisfying $|S| = |G|$, then $\psi(S) < \psi(G)$.

Theorem 5.2.5 If $|G| = n$ then $P(G) \leq P(C_n)$, with equality if and only if $G \cong C_n$.

Theorem 5.2.7 If $|G| = n$ and $r < 0$ is a real number, then $R_G(r) \geq R_{C_n}(r)$, with equality if and only if G is nilpotent.

Question 5.2.10 Are there reals (r, s) such that all finite groups G of order n satisfying $R_G(r, s) = R_{C_n}(r, s)$ are soluble?

Corollary 5.3.2 If G is a nilpotent group of order n and r is a real number, then $R_G(r) = R_{C_n}(r)$. In particular, $R_G(1) = R_{C_n}(1)$.

Conjecture 5.3.3 If $|G| = n$ and $R_G(1) = R_{C_n}(1)$, then G is nilpotent.

Conjecture 5.3.4 $R_G(1) \leq R_{C_n}(1)$ for all finite groups G of order n.

The authors would like to thank the anonymous referees of this paper for many helpful comments and useful suggestions.

2 Properties of Periodic Groups G Determined by $\omega(G)$

In this section, G denotes a periodic group. Recall that $\omega(G)$ denotes the set of orders of all elements of G, the so-called *spectrum* of G.

2.1 Some Known Results About the Function $\omega(G)$

It is well known that $\omega(G) = \{1, 2\}$ if and only if G is an elementary abelian 2-group. The following results are more complicated and their proofs may be found in the quoted literature.

Theorems 2.1.1 (1) *If* $\omega(G) = \{1, 3\}$, *then* G *is nilpotent of class* ≤ 3, *(F.Levi and B.L. van der Waerden,* 1933, *(see [33])).*

(2) *If* $\omega(G) = \{1, 2, 3\}$, *then* G *is (elementary abelian)-by-(prime order), (B.H. Neumann,* 1937, *(see [52])).*

(3) *If* $\omega(G) \subseteq \{1, 2, 3, 4\}$, *then* G *is locally finite (I.N. Sanov,* 1940, *(see [57])).*

(4) *If* $\omega(G) = \{1, 2, 3, 6\}$, *then* G *is locally finite (M. Hall,* 1958, *(see [14])).*

(5) *If* $\omega(G) = \{1, 2, 3, 4, 6\}$, *then* G *is locally finite (A.S. Mamontov,* 2013, *(see [40])).*

(6) *If* $\omega(G) = \{1, 2, 5\}$, *then* G *is locally finite (M.F. Newman* 1979, *E. Jabara* 2004, *(see [53] and [23])).*

(7) *If* $\omega(G) = \{1, 2, 3, 5, 6\}$, *then* G *is soluble and locally finite (V.D. Mazurov, A.S. Mamontov* 2009, *(see [41])).*

(8) *If* $\omega(G) = \{1, 2, 3, 4, 5, 6\}$, *then* G *is locally finite (D.V. Lytkina, V.D. Mazurov, A.S. Mamontov and E. Jabara,* 2014, *(see [27])).*

(9) *If* $\omega(G) \subseteq \{1, 2, 3, 4, 5, 6\}$ *and* $\omega(G) \neq \{1, 5\}$, *then* G *is locally finite (E. Jabara, D.V. Lytkina, V.D. Mazurov and A.S. Mamontov,* 2014, *(see [27])).*

(10) *If* $\omega(G) = \{1, 2, 3, 6, 7\}$, *then either* G *is locally finite or an extension of a non-trivial abelian 2-group by a group without involutions (W. Guo, A.S. Mamontov,* 2017 *(see [12])).*

(11) *If* $\omega(G) = \{1, 2, 3, 4, 8\}$, *then* G *is locally finite (V.D. Mazurov,* 2011, *(see [44])).*

(12) *If* $\omega(G) = \{1, 2, 3, 4, 9\}$, *then* G *is locally finite (E. Jabara, D. Lytkina,* 2013, *(see [24])).*

(13) *If* $\omega(G) = \{1, 2, 3, 4, 8, 9\}$, *then* G *is locally finite (E. Jabara, D. Lytkina, V.D. Mazurov* 2014, *(see [25])).*

(14) *If* $\omega(G) = \{1, 2, 3, 4, p, 9\}$, *where* $p \in \{5, 7\}$, *then* G *is locally finite (E. Jabara, A.S. Mamontov,* 2016, *(see [29])).*

2.2 Some Open Questions About the Function $\omega(G)$

There are also many open questions related to the function $\omega(G)$.

Questions 2.2.1 *Does the following assumption imply that* G *is locally finite?*

(1) $\omega(G) = \{1, 5\}$;
(2) $\omega(G) = \{1, 7\}$;
(3) $\omega(G) = \{1, 11\}$;
(4) $\omega(G) = \{1, 2, 5, 10\}$;
(5) $\omega(G) = \{1, 2, 4, 8\}$;
(6) $\omega(G) = \{1, 2, 3, 4, 5, 6, 7\}$;
(7) $\omega(G) = \{1, 2, 3, 4, 8, 16\}$.

2.3 More Precise Results

In some cases, there are more precise results on the structure of G.

Theorems 2.3.1 (1) *If $\omega(G) = \{1, 2, 5\}$, then G is either an extension of an elementary abelian 2-group A by a group P of order 5 acting without fixed points on A, or an extension of an elementary abelian 5-group P by a group of order 2 acting without fixed points on P (M.F. Newman, 1979, E. Jabara, 2004, (see [53] and [23])).*

(2) *If $\omega(G) = \{1, 2, 3, 4\}$, then G is either (elementary abelian 3-group)-by-(cyclic or quaternion group) or (nilpotent 2-group of class 2)-by-(a subgroup of S_3) (D.V. Lytkina, 2007, (see [39])).*

(3) *If $\omega(G) = \{1, 3, 4, 7\}$, then $G \cong L_2(7)$ (A.A. Kuznetsov and D. Lytkina, 2007, (see [34])).*

(4) *If $\omega(G) = \{1, 2, 3, 4, 5, 8\}$, then $G \cong M_{10}$ (E. Jabara, D. Lytkina and A.S. Mamontov, 2013, (see [26])).*

(5) *If $\omega(G) = \{1, 2, 3, 4, 5, 7\}$, then $G \simeq L_3(4)$ (E. Jabara and A.S. Mamontov, 2015, (see [28])).*

(6) *If G is a finite group and $\omega(G) = \{1, 2, 3, 4, 5, 6, 7\}$, then $G \simeq A_7$ (R. Brandl and W.J. Shi, 1991, (see [3])).*

(7) *If $\omega(G) = \{1, 3, 5\}$, then either $G = FT$ where F is a normal 5-subgroup which is nilpotent of class 2 and $|T| = 3$, or F is a normal 3-subgroup which is nilpotent of class 3 and $|T| = 5$ (N.D. Gupta and V.D. Mazurov, 1999, (see [13])).*

(8) *If $\omega(G) = \{1, 2, 4, 5\}$, then $G = TD$ where T is an elementary abelian 2-group and D is non-abelian of order 10, or $G = FT$, where F is an elementary abelian normal 5-subgroup and T is isomorphic to a subgroup of Q_8, or $G = TF$, where T is a nilpotent normal 2-subgroup of class at most 6 and $|F| = 5$ (N.D. Gupta and V.D. Mazurov, 1999, and E. Jabara, 2004, (see [13] and [23])).*

(9) *If $\omega(G) = \{1, 2, 3, 5, 6\}$, then G is an extension of an elementary abelian 5-group by a cyclic group of order 6, or G is an extension of a 3-group of class at most 3 by the dihedral group D_{10}, or G is an extension of the direct product of a 3-group of class at most 2 and an elementary abelian 2-group by a group of order 5 (A.S. Mamontov and V.D. Mazurov, 2009, (see [41])).*

(10) *If $\omega(G) = \{1, 2, 3, 4, 5\}$, then either $G \simeq A_6$ or $G = VC$, where V is a nontrivial elementary abelian normal 2-subgroup and $C \simeq A_5$ (V.D. Mazurov, 2000, (see [45])).*

(11) *If $\omega(G) = \{1, 2, 3, 4, 8\}$, then either (i) $G = VQ$ where V is a nontrivial normal abelian 3-subgroup and Q is a 2-subgroup acting without fixed points on V and either cyclic of order 8 or isomorphic to the quaternion group Q_{16}, or (ii) $G = T\langle a\rangle$ where T is a normal 2-subgroup nilpotent of class at most 2 and exponent 8, $|a| = 3$ and a acts without fixed points on T, or (iii) $G = TS$ where T is a normal 2-subgroup which is nilpotent of class at most 2 and exponent 4 and $S \simeq S_3$ acts without fixed points on T (V.D. Mazurov 2011, (see [44])).*

It is still an open question if for every group G with $\omega(G) = \{1, 2, 3, 4, 5, 6, 7\}$ we have $G \cong A_7$.

The reader can consult [22, 35–38, 41, 45, 46] for more results.

2.4 A Sketch of the Proof of a Theorem of Mazurov

We give a sketch of the proof of Theorem 2.3.1(11) due to V.D. Mazurov (see [44])). As for many other mentioned theorems, the proof will use methods of local analysis as well as calculations in GAP.

Theorem 2.4.1 *Let G be a group with $\omega(G) = \{1, 2, 3, 4, 8\}$. Then G is locally finite and one of the following holds:*

(i) *$G = VQ$ where V is a nontrivial normal abelian 3-subgroup and Q is a 2-subgroup acting freely on V and either cyclic of order 8 or isomorphic to the quaternion group Q_{16},*

(ii) *$G = T\langle a\rangle$ where T is a normal 2-subgroup nilpotent of class at most 2 and exponent 8, $|a| = 3$ and a acts without fixed points on T,*

(iii) *$G = TS$ where T is a normal 2-subgroup nilpotent of class at most 2 and exponent 4 and $S \cong S_3$ acts without fixed points on T.*

Sketch of the Proof Let $R = \langle r\rangle$ be a Sylow 3-subgroup of G.

First assume that, for every involution t of G, the product rt is an involution. We claim that in this case G has the structure in (*i*).

Let $t \in G$ be an involution. The subgroup $V = C_G(r)$ is a 3-group of exponent 3, by the hypothesis on $\omega(G)$. We have $(rt)^2 = 1$; hence, $r^t = r^{-1}$, and thus $\langle r\rangle$ is t-invariant and then V is t-invariant. Since $C_V(t) = 1$, t acts on V as a fixed-point-free automorphism of order 2; hence, V is abelian and then it is elementary abelian. If $y \in G$ is an involution, then $r^y = r^{-1}$, and hence $ty \in V$. Therefore, $V\langle t\rangle$ contains all involutions of G. For every $x \in V$, we have $(xt)^2 = 1$ and $x = xt \cdot t$; therefore, the subgroup $V\langle t\rangle$ is generated by all involutions, and hence it is normal in G. It follows that V is normal in G. Now, if $g \in G$, then we have $t^g \in V\langle t\rangle$, $a = tt^g \in V$ and $t^{ga} = a^{-1}t^g a = a^{-2}t^g = at^g = tt^g \cdot t^g = t$, therefore $ga \in C_G(t)$. Hence, $G = VQ$ where $Q = C_G(t)$. Obviously, Q is a 2-group, and the group VQ is a Frobenius group since G has no elements of order 6, and thus t is the unique involution in Q. Hence, the group $Q/\langle t\rangle$ is a group of exponent 4. By Sanov theorem, Q is locally finite, and hence either is cyclic of order 8, or it is isomorphic to a generalized quaternion group of order 16. Therefore, G has the structure in (*i*).

Now assume that there exists an involution $y \in G$ such that ry has order 3, 4, or 8.

We show that, in this case, there exists a subgroup H of G containing R which is either an extension of a nontrivial 2-group by a group of order 3 (we say in this case that H is a subgroup of type 1), or an extension of a nontrivial 2-group by a group which is isomorphic to S_3 (we say in this case that H a subgroup of type 2).

In fact, if ry has order 3, then from $ryry = y^{-1}r^{-1}$ we get $r^2yry = ry^{-1}r^{-1}$; hence, the element y^ry has order 2, $yy^r = y^ry$ and the group $\langle y, r\rangle = \langle y, y^r\rangle\langle r\rangle \simeq A_4$ is of type 1.

Now suppose that ry has order ≥ 4. Then $(ry)^2 = ryry$ is a product of two elements, r, r^y, of order 3, such that $(rr^y)^4 = 1$. Calculation with GAP shows that, if $k, l, m, n \in \{3, 8\}$, the group $\langle a, b \mid a^3 = b^3 = (ab)^4 = (ab^{-1})^8 = (abab^{-1})^k = (baba^{-1})^l = (bab^{-1}a^{-1})^m = (abab^{-1}a^{-1}b^{-1})^n\rangle$ is trivial. It follows that $(rr^{-y})^3 = 1$; hence, $1 = rr^yr^yrr^{-y}r^{-1}r^{-1}r^{-y}$, thus $rr^yr^yr = r^yrrr^y$, thus $A = \langle rr^y, r^yr\rangle$ is an abelian 2-group. Moreover, $(rr^y)^y = r^yr$, $(r^yr)^y = rr^y$, $(rr^y)^r = r^yr$, $(r^yr)^r = r^{-1}r^yr^{-1} = (r^yr)^{-1}(rr^y)^{-1}$. Therefore, the subgroup A is normal in $F = \langle r, y\rangle$. Furthermore, the element r^yr^{-1} has order 3, and we have $(r^yr^{-1})^y = rr^{-y}$, thus the subgroup $S = \langle r^yr^{-1}, y\rangle \simeq S_3$. Therefore, the subgroup $F = AS$ is finite of type 2.

Now the proof of the theorem follows from the following two results, whose proofs are quite intricate, and therefore we refer for them to the original paper.

(a) If there exists a subgroup of G of type 1 which contains R and if there does not exist a subgroup of type 2 which contains R, then G has the structure in *(ii)*.
(b) If there exists a subgroup of G of type 2 which contains R, then G has the structure in *(iii)*. □

2.5 Some More Questions and Results

A very well-known question is if G is locally finite, whenever $\omega(G)$ is finite; this is the famous *Burnside problem*, which was answered negatively by Novikov and Adjan in 1968 (see [54]). Another related question is the following: if it is known that a group G with m generators and exponent n is finite, can one conclude that the order of G is bounded by some constant depending only on m and n? Equivalently, are there only finitely many finite groups with m generators of exponent n, up to isomorphism? This is the *Restricted Burnside problem*. Efim Zelmanov showed in [65, 66] that the answer is affirmative. He was awarded the Fields Medal in 1994 for his work.

Another very famous problem related to $\omega(G)$ is the so-called *recognizability by spectrum*. A periodic group G is said to be *recognizable by spectrum in a class $\mathcal{C}$* if, for every periodic group $H \in \mathcal{C}$, from $\omega(G) = \omega(H)$ it follows $G \cong H$. For example, as we mentioned before, $\omega(G) = \omega(L_2(7))$ implies $G \cong L_2(7)$ (see [34]), $\omega(G) = \omega(M_{10})$ implies $G \cong M_{10}$ (see [26]) and $\omega(G) = \omega(L_3(4))$ implies $G \cong L_3(4)$ (see [28]); hence, $L_2(7), M_{10}, L_3(4)$ are recognizable by spectrum in the class of all groups. On the other hand, V.D. Mazurov and W.J. Shi in 2012 proved the following result (see [47]):

Theorem 2.5.1 *Let G be a finite group. If G contains a nontrivial normal soluble subgroup, then there exist infinitely many non-isomorphic finite groups X with $\omega(G) = \omega(X)$. Conversely, if there exist infinitely many non-isomorphic finite groups X with $\omega(G) = \omega(X)$, then there exists a group S, with $\omega(S) = \omega(G)$, which contains a nontrivial normal soluble subgroup.*

Proof Assume that G contains a nontrivial normal subgroup, then G contains a nontrivial normal subgroup V which is an elementary abelian p-group for some prime p. Let G_1 be the semidirect product of V and G, where the action of G on V is the conjugation in G. Write $G_1 = \{(g, v) \mid g \in G, v \in V\}$. We claim that $\omega(G_1) = \omega(G)$.

In fact, obviously $\omega(G) \subseteq \omega(G_1)$. Let $(g, v) \in G_1$ and write m the order of (g, v). Let n be the order of g in G/V. Then $g^n \in V$, write $u = g^n$. If $u \neq 1$, then the order of g is np, but from $(g, v)^n = (u, w)$ with $w \in V$ we get that the order of $(g, v)^n$ is equal to p, therefore $m = np = o(g) \in \omega(G)$. Now suppose $u = 1$, then $o(g) = n$ and assume $o(g, v) \neq n$. Then $1 \neq (g, v)^n = (1, vv^g \cdots v^{g^{n-1}})$; hence, $(gv)^n = vv^g \cdots v^{g^{n-1}} \neq 1$. Therefore, the element gv of G has order np and $m = np \in \omega(G)$.

Therefore, $\omega(G) = \omega(G_1)$ and obviously $|G| < |G_1|$.

The group G_1 contains a nontrivial normal abelian subgroup and we can repeat the process to construct a finite group G_2 with $\omega(G_2) = \omega(G_1) = \omega(G)$, with a nontrivial normal abelian subgroup. By induction, it is possible to construct infinitely many finite non-isomorphic groups G_n with $\omega(G_n) = \omega(G)$.

In order to prove the converse, first we show the following result.

(a) For every finite set ω of natural numbers, there exists a natural number $k = k(\omega)$ such that every group whose spectrum coincides with ω contains a subgroup X with $\omega(X) = \omega$ and $|X| < k$.

Let $\omega = \{n_1, \cdots, n_s\}$, $n = n_1 \cdots, n_s$, and x_i be some element of order n_i in G, with $i = 1, \cdots, s$. Suppose $X = \langle x_1, \cdots, x_s \rangle$. Then $\omega(S) = \omega$ and $x^n = 1$ for any $x \in X$. Now, the positive solution to the restricted Burnside problem (see [Z1] and [Z2]) implies that there exists a number $k = k(s, n)$ for which $|X| < k$, and obviously s and n depend only on ω.

Now, suppose that there exist infinitely many pairwise non-isomorphic groups T with $\omega(T) = \omega(G)$, none of which contains a nontrivial soluble normal subgroup. Let H be such a group and denote by D the socle of H. Then $D = P_1 \times \cdots \times P_r$, where $P_1, \cdots, P_s$ are non-abelian simple groups and H acts on $\Delta = \{P_1, \cdots, P_r\}$ by conjugation. Write $k = k(\omega(G))$ the integer defined in (a). We show that

(b) $r \leq k^2$.

Assume $r > k^2$. Let M be a subgroup of H such that $\omega(M) = \omega(H)$ and $|M| < k$. Then M acts on Δ by conjugation, and, since the length of each orbit is at most k and $r > k^2$, the group D is a direct product of at least k nontrivial normal subgroups $N_1, \cdots, N_t$ of DM. Since the subgroups N_i, $i \in \{1, \cdots, t\}$ have pairwise trivial intersection and $t \geq k > |M|$, there exists i for which $N_i \cap M = 1$. Let P be a nontrivial Sylow subgroup of N_i. By the Frattini's argument, $N_iM = N_iN_{N_iM}(P)$; hence, $M \cong N_iM/N_i \cong (N_{N_iM}(P))/(N_{N_iM}(P) \cap N_i)$. Since $\omega(M) = \omega(G)$, we have

$\omega(N_{N_iM}(P)) = \omega(G)$ and the group $N_{N_iM}(P)$ contains the normal soluble subgroup P, a contradiction. Therefore, $r \leq k^2$ and (b) is proved.

Now let π be the set of primes dividing the elements of $\omega(G)$. From the classification of finite simple groups, it follows that there exists a finite number of non-isomorphic simple groups S with $\pi(|S|) \subseteq \pi$. Then, there exists a positive integer s such that $|P_i| \leq s$ for every $i \in \{1, \cdots, r\}$ and $|D| \leq s^{k^2}$, since $r \leq k^2$. Since $C_H(D) = \{1\}$, we obtain H isomorphic to a subgroup of $Aut(D)$; hence, $|H| \leq |Aut(D)| \leq (s^{k^2})!$. Therefore, $|H|$ is bounded by a function of $\omega(G)$, which is in contradiction with the assumption that there exist infinitely many such groups H.

The theorem is proved. □

Theorem 2.5.1 raises several questions. First of all you may ask whether a finite group and a simple group must be isomorphic if they have the same set of orders of elements. The answer is NO, since it is possible to see that $\omega(L_3(5)) = \omega(L_3(5) : 2)$. In fact, V.D. Mazurov in 1994, in the paper [43], proved that the only finite group with $\omega(G) = \omega(L_3(5))$ except $L_3(5)$ is $(L_3(5) : 2)$. This result also gives a negative answer to Question 12.84 in the Kourovka Notebook (see [31]).

There is a large literature dealing with the recognizability problems for finite simple and almost simple groups, in the class of all groups and in the class of all finite groups. For example, if G is either a Suzuki simple group or $G = L_2(q)$ with $q \neq 9$, and H is a finite group, then $\omega(G) = \omega(H)$ implies $G \cong H$ (see, resp., [3, 4]), while it is possible to construct a finite group H with $\omega(H) = \omega((L_2(9))$ and H non-isomorphic to $L_2(9)$ (see [3]). Some more examples of finite simple groups recognizable by spectrum in the class of finite groups are the alternating groups A_n, where $n \neq 6, 10$, as I.B. Gorshkov proved in [11].

Notice that there are finite groups that are recognizable by spectrum in the class of finite groups but not in the class of all groups; an example is the group $L_2(2^{61} - 1)$, as V.D. Mazurov, A.Yu Ol'shanskii and A.I. Sozutov proved in [48].

Another very interesting question was raised by C. Praeger and W.J. Shi in [55] and by W.J. Shi in [59], which is also Question 12.39 in the Kourovka Notebook (see [31]).

Question 2.5.2 *Must a finite group and a finite simple group be isomorphic if they have equal orders and the same set of orders of elements?*

The answer is YES, as the following identification theorem of A.V. Vasil'ev, M.A. Grechkoseeva, and V.D. Mazurov shows (see [60]).

Theorem 2.5.3 *If S is a simple finite group and G is a finite group satisfying $|G| = |S|$ and $\omega(G) = \omega(S)$, then $G \cong S$.*

We end this section with the following result by A.A. Buturlakin (see [5]), which answers a question posed by A.V. Valisev (see Problem 16.25 in [31]).

Theorem 2.5.4 *Three pairwise non-isomorphic finite non-abelian simple groups with the same spectrum do not exist.*

3 Roots of Unity in Finite Groups

In this section, G denotes a finite group.

Definition 3.1 Let e be a divisor of $|G|$. Then

$$L_e(G) := \{x \in G : x^e = 1\}.$$

Thus $L_e(G)$ is the set of e-th **roots of unity** in G.

We shall deal with the following problem:

How do the sizes of the subsets $L_e(G)$ affect the structure of G?

3.1 A Result of Frobenius

The best known result in this direction is the following theorem of G. Frobenius from 1895 (see [9] and Chapter 9 in [15]).

Theorem 3.1.2 *For every e dividing $|G|$, we have that e divides $|L_e(G)|$.*

Frobenius also made the following conjecture:

Conjecture 3.1.3 *If e divides $|G|$ and $|L_e(G)| = e$, then $L_e(G)$ is a characteristic subgroup of G.*

M. Hall proved the conjecture when G is soluble (see [15]).

Proposition 3.1.3 *Let G be a soluble group of order n and let e be a divisor of n such that the equation $x^e = 1$ has exactly e solutions, then these solutions form a characteristic subgroup of G.*

Proof Obviously, it is enough to show that $L_e(G)$ is a subgroup of G. If e is a prime, the result is obviously true. By induction assume the result true for soluble groups of order less than $|G|$. Let K be a minimal normal subgroup of G. Then, K is an elementary abelian p-group, where p is a prime. Write $|K| = p^i$. We consider two cases, the case that p divides e and the case that $(p, e) = 1$.

Case 1: p divides e. In this case, $x^e = 1$ for every $x \in K$, thus $K \subseteq L_e(G)$. Let $e = p^j e_1, n = p^s n_1$ with $(e_1, p) = 1 = (n_1, p)$. Then, G/K has order $p^{s-i}n_1$ and has order divisible by $u = p^{j-i}e_1$ if $j \geq i$, $u = e_1$ if $j < i$. By Frobenius' theorem, there exists a positive integer k such that there are ku elements z such that $z^u = 1$ in G/K. Now we show that if $z = xK$ and $z^u = 1$, then $x^e = 1$. In fact, from $(xH)^u = H$ we get $x^u \in K$ and then $x^{up} = 1$, but up divides e, and thus $x^e = 1$. Therefore, the elements of ku cosets in G/K are in $L_e(G)$ and then there are at least kup^i elements in $L_e(G)$. Now, if $j < i$, then $up^i > e$ and we obtain a contradiction since $|L_e(G)| = e$. Hence, $j \geq i$, then $e = up^i$ and there are ke elements in $L_e(G)$. Therefore, $k = 1$. Then in

G/K there are u elements z such that $z^u = 1$. By induction, these elements form a normal subgroup H/K of order u. Therefore, H is a normal subgroup of G of order $e = p^i u$ and its elements are exactly the e solutions of $x^e = 1$.

Case 2: p does not divide e. In this case, e divides $|G/K|$ and, by Frobenius' theorem, there exists a positive integer k such that in G/K there are ke elements z such that $z^e = 1$. If $z = yK$ and $z^e = 1$, then $(yK)^e = K$ and $y^e \in K$, and thus $y^{ep} = 1$. Hence, in G/K there are ke cosets consisting of elements y satisfying $y^{ep} = 1$. Now we show that each coset yK such that $(yK)^e = K$ yields a solution of $x^e = 1$, and that different cosets solutions of $(yK)^e = K$ yield different solutions of $x^e = 1$. In fact, if $(yK)^e = K$, then $y^{ep} = 1$ and $x = y^p$ satisfies $x^e = 1$; moreover, if y_1K, y_2K are different and such that $(y_1K)^e = K = (y_2K)^e$, then $x_1 = y_1^p$, $x_2 = y_2^p$ are different otherwise from $y_1^pK = y_2^pK$ and $(p, e) = 1$ we would have $y_1K = y_2K$. Therefore, if there are ke solutions of $z^e = 1$ in G/K, we have at least ke solutions of $x^e = 1$. Thus, $k = 1$ and by induction the solutions of $z^e = 1$ in G/K form a subgroup U/K. Then, the group U has order $p^i e$ with $(p, e) = 1$. But U is soluble; hence, by Hall's Theorem, there exists a subgroup of U, say H, of order e. Then the elements h of H satisfy $h^e = 1$ and are exactly the elements of $L_e(G)$. Therefore, $L_e(G)$ is a subgroup and we have the result. □

The full conjecture was proved in 1991 by N. Iiyori and H. Yamaki (see [21]).

Theorem 3.1.4 *Let G be a finite group of order n and let e be a divisor of n such that the equation $x^e = 1$ has exactly e solutions. Then, these solutions form a characteristic subgroup of G.*

We omit the proof that uses the classification of finite simple groups and some previous results by H. Yamaki (see [61–64]).

3.2 More Results Concerning $|L_e(G)|$

It is easy to see that the following result holds:

Proposition 3.2 $|L_e(G)| = e$, *for every e dividing $|G|$, if and only if G is cyclic.*

Proof If G is a finite cyclic group, then for every e dividing $|G|$ there exists exactly one subgroup H of G of order e, and we have $H = \{x \in G \mid x^e = 1\}$.

Conversely, suppose that $|L_e(G)| = e$ for every e dividing $|G|$. First, we show that every Sylow p-subgroup P of G is cyclic, for every prime p. In fact, if $|P| = p^n$ and $exp(P) \leq p^{n-1}$, then $P \subseteq \{x \in G \mid x^{p^{n-1}} = 1\}$, and thus $p^n = |P| \leq p^{n-1}$, a contradiction. Therefore $exp(P) = p^n$. Moreover, from $P \subseteq \{x \in G \mid x^{p^n} = 1\}$ which is of order p^n, it follows $P = \{x \in G \mid x^{p^n} = 1\}$. Hence, every Sylow p-subgroup is also normal in G. Therefore, G is the direct product of its Sylow p-subgroups, and then it is cyclic. □

In 2011, W. Meng and J. Shi (see [50]) introduced the following problem.
Inverse Problem to Frobenius' Theorem *For a small positive integer k, give a complete classification of all groups G with $|L_e(G)| \leq ke$ for every e.*

If m is a positive integer, denote by $Div(m)$ the set of all divisors of m. The following function between positive integers is defined in [16] and in [50]:

$$b : e \in Div(|G|) \longmapsto b(e) = |L_e(G)| \in \mathbb{N}.$$

For example, if $G \cong S_3$, then $b(2) = 4$, $b(3) = 3$, $b(6) = 6$; if $G \cong Q_8$, then $b(2) = 2$, $b(4) = 8$.

H. Heineken and F. Russo introduced in [16] the following generalization of the previous problem.
Generalized Inverse Problem to Frobenius' Theorem *Classify all finite groups G such that $b(e) \leq f(e)$ for all $e \in Div(exp(G))$, where $f : e \in Div(|G|) \longmapsto f(e) \in \mathbb{N}$ is a prescribed arithmetic function depending only on e.*

Meng and Shi in [50] classified all groups G satisfying

$$|L_e(G)| \leq 2e \quad \text{for every } e \text{ dividing } |G|,$$

i.e., they solved the generalized inverse problem if $f(e) = 2e$.

They proved the following theorem.

Theorem 3.2.1 *Let G be a finite group. Then $b(e) \leq 2e$ for all e dividing $|G|$ if and only if one of the following statements holds:*

(1) G is cyclic;
(2) $G \cong C_m \times C_{2^{k-1}} \times C_2$ with m odd and $k \geq 2$;
(3) $G \cong C_m \times Q_8$ with m odd;
(4) $G \cong C_m \times SD_{2^t}$ where $SD_{2^t} = \langle a, b \mid a^{2^{t-1}} = b^2 = 1, b^{-1}ab = a^{1+2^{t-2}} \rangle$ with $t \geq 4$ is the semidihedral 2-group of order 2^t and m odd;
(5) $G \cong C_m \times \langle a, b \mid a^3 = b^{2^s} = 1, b^{-1}ab = a^{-1} \rangle$ with $s \geq 1$ and $(m, 6) = 1$.

Corollary 3.2.2 *Let G be a finite group. Then $b(e) = 2e$ for all e dividing $|G|$ if and only if one of the following statements holds:*

(1) $G \cong C_{2^{k-1}} \times C_2$ where $k \geq 2$;
(2) $G \cong SD_{2^t}$ where $SD_{2^t} = \langle a, b \mid a^{2^{t-1}} = b^2 = 1, b^{-1}ab = a^{1+2^{t-2}} \rangle$ with $t \geq 4$ is the semidihedral 2-group of order 2^t.

In 2015, H. Heineken and F. Russo (see [16]) described groups G satisfying

$$|L_e(G)| \leq e^2 \quad \text{for every } e \text{ dividing } |G|.$$

Among many other results, they proved that all groups with this property are soluble and of fitting length at most 3. In the paper [17], H. Heineken and F. Russo introduced the symbol $B(G)$, **the global breath of G in the sense of Frobenius**, or, briefly, **the global breath of G**, as follows:

$$B(G) = max\left\{\frac{|L_e(G)|}{e} \mid e \in Div(exp(G))\right\}.$$

For example, $B(G) = 1$ if and only if G is cyclic, $B(Q_8) = 2$, $B(D_{2^n}) = 2^{n-2} + 1$. Theorem 3.2.1 describes groups G with $B(G) = 2$. Groups G with $B(G) = 3$ have been completely described by W. Meng, J. Shi and K. Chen in [51]. For example, if G is a p-group, p a prime, they proved the following result.

Theorem 3.2.3 *Let G be a group of order p^n, p a prime. Then, $B(G) = 3$ if and only if one of the following holds:*

(1) (i) $G \cong C_{3^{k-1}} \times C_3$ where $k \geq 2$;
(2) (ii) $G \cong \langle a, b \mid a^{3^{t-1}} = b^3 = 1, b^{-1}ab = a^{1+3^{t-2}}\rangle$ where $t \geq 3$;
(3) (iii) $G \cong Q_{2^4}$, the generalized quaternion group of order 16;
(4) (iv) $G \cong D_8$, the dihedral group of order 8;
(5) (v) $G \cong QD_{2^4} = \langle a, b \mid a^8 = b^2 = 1, b^{-1}ab = a^3\rangle$.

Recently, W. Meng in [49] investigated the group G with $B(G) = 4$. Among many other results, he proved that a group with this property is soluble and of even order.

If $G \cong A_4 \times C_2$, then $b(2) = b(4) = b(8) = 8$, $b(3) = 9$ and $b(6) = b(12) = b(24) = 24$. Therefore, $B(A_4 \times C_2) = 4$, and thus there exists a non-supersolvable group G with $B(G) = 4$ (see [49]).

We end this section with the following question posed by J.G. Thompson, which is Question 12.37 in the Kourovka Notebook (see [31]):

Question 3.2.4 *Let G be a soluble group. Is it true that if H is a finite group satisfying $|L_e(H)| = |L_e(G)|$ for every $e \in \mathbb{N}$, then also H is soluble?*

This question is still open.

4 Results Concerning the Function $\psi(G)$

4.1 Basic Facts About the Function $\psi(G)$

Recall that G denotes a finite group, p denotes a prime number and

$$\psi(G) := \sum_{x \in G} o(x).$$

We shall deal with the following problem:

How does the value of $\psi(G)$ affects the structure of G?

In order to get acquainted with the function $\psi(G)$, we shall present its values for some small groups and for the cyclic p-groups. First notice that

$$\psi(C_p) = 1 + (p - 1)p,$$

so

$$\psi(C_2) = 3, \quad \psi(C_3) = 7, \quad \psi(C_5) = 21 \quad \text{and} \quad \psi(C_7) = 43.$$

Moreover, it is easy to check that

$$\psi(C_4) = 1 + 2 + 2 \cdot 4 = 11 \text{ and } \psi(C_6) = 1 + 2 + 2 \cdot 3 + 2 \cdot 6 = 21.$$

For groups of exponent p, we have $\psi(G) = p|G| - (p-1)$. In particular, $\psi(C_2 \times C_2) = 7$. Finitely, we mention that

$$\psi(S_3) = 1 + 3 \cdot 2 + 2 \cdot 3 = 13.$$

As shown above, $\psi(C_5) = 21 = \psi(C_6)$. Therefore, the value of $\psi(G)$ does not determine the group G. Moreover, the values of $\psi(G)$ and $|G|$ still do not determine the group G. Indeed, let $A = C_8 \times C_2$ and $B = C_8 \rtimes C_2$, where $C_8 = \langle a \rangle$, $C_2 = \langle b \rangle$ and $a^b = a^5$. Then, $|A| = |B| = 16$ and it can be shown that $\psi(A) = \psi(B) = 87$.

Notice, however, that S_3 is the only group satisfying $\psi(G) = \psi(S_3) = 13$.

In the following theorem, we collect some simple, but important, results concerning the function $\psi(G)$.

Theorem 4.1.1 *Let r denote a positive integer. Then, the following statements hold.*

(1) If $G = A \times B$ for some subgroups of G of coprime orders, then $\psi(G) = \psi(A)\psi(B)$.

(2) If P is a cyclic group of order p^r, then

$$\psi(P) = \frac{p^{2r+1}+1}{p+1} = \frac{p|P|^2+1}{p+1} > \frac{p}{p+1}|P|^2.$$

(3) Let $q = p_1 < p_2 < \cdots < p_t = p$ be the prime divisors of the integer n and let the corresponding Sylow subgroups of C_n be $P_1, P_2, \ldots, P_t$. Then

$$\psi(C_n) = \prod_{i=1}^{t} \psi(P_i) \geq \frac{q}{p+1} n^2 \geq \frac{2}{p+1} n^2.$$

Proof (1) It is easy to see that if $G = A \times B$ and $(|A|, |B|) = 1$, then $\psi(G) = \psi(A)\psi(B)$.

(2) If P is a cyclic group of order p^r, then
$\psi(P) = 1 + p\varphi(p) + p^2\varphi(p^2) + \cdots + p^r\varphi(p^r) = 1 + (p-1)(p + p^3 + p^5 + \cdots + p^{2r-1}) =$

$$\frac{p^{2r+1}+1}{p+1} = \frac{p|P|^2+1}{p+1} > \frac{p}{p+1}|P|^2,$$

where φ denotes the Euler's function.

(3) Since $C_n = P_1 \times P_2 \times \cdots \times P_t$, it follows by (1) that $\psi(C_n) = \prod_{i=1}^{t} \psi(P_i)$. Since $p_{i+1} \geq p_i + 1$ for all i and $p_1 = q \geq 2$, it follows by (2) that

$$\psi(C_n) > \prod_{i=1}^{t} \frac{p_i}{p_i+1}|P_i|^2 \geq \frac{p_1}{p_t+1}n^2 = \frac{q}{p+1}n^2 \geq \frac{2}{p+1}n^2.$$

□

4.2 The Theorem of Amiri, Jafarian Amiri, and Isaacs

The notation $\psi(G)$ was introduced by H. Amiri, S.M. Jafarian Amiri, and I.M. Isaacs in their paper [1] from 2009. In that paper, they proved the following basic theorem.

Theorem 4.2.1 *If G is a noncyclic finite group of order n and if C_n denotes the cyclic group of order n, then*

$$\psi(G) < \psi(C_n).$$

Recently, R. Shen, G. Chen, and C. Wu in [58] and S.M. Jafarian Amiri and M. Amiri in [30] studied noncyclic finite groups G of order n with the largest value of $\psi(G)$, and obtained information about the structure of such groups in certain cases.

4.3 The Main Results of Herzog, Longobardi, and Maj

In a recent paper (see [18]), the authors of this survey continued to study the function $\psi(G)$. Our main result determines the best possible upper bound for the value of $\psi(G)$ for noncyclic groups of order n. We proved the following theorem.

Theorem 4.3.1 *If G is a noncyclic finite group of order n, then*

$$\psi(G) \leq \frac{7}{11}\psi(C_n).$$

This upper bound is the best possible. Indeed, as shown in the following proposition, for each $n = 4k$, where k denotes an odd integer, there exists a noncyclic group G of order n satisfying $\psi(G) = \frac{7}{11}\psi(C_n)$.

Proposition 4.3.2 *Let k be an odd integer and let $n = 4k$. Then, $|C_{2k} \times C_2| = n$ and*

$$\psi(C_{2k} \times C_2) = \frac{7}{11}\psi(C_n).$$

Proof Since $n = 4k$ and $(4, k) = 1$, we have

$$\psi(C_n) = \psi(C_4 \times C_k) = \psi(C_4)\psi(C_k) = 11\psi(C_k).$$

Thus, $\psi(C_k) = \frac{1}{11}\psi(C_n)$.

Next notice that $C_{2k} \times C_2 = C_k \times C_2 \times C_2$. Hence,

$$\psi(C_{2k} \times C_2) = \psi(C_2 \times C_2)\psi(C_k) = 7 \cdot \frac{1}{11}\psi(C_n),$$

as required. □

Our second major result is the following theorem.

Theorem 4.3.3 *Let G be a noncyclic group of order n and let q be the smallest prime divisor of n. Then,*

$$\psi(G) < \frac{1}{q-1}\psi(C_n).$$

For groups of even order $q = 2$ and Theorem 4.3.3 only implies that $\psi(G) < \psi(C_n)$, as shown already in Theorem 4.2.1. But for groups of odd order $q \geq 3$ and Theorem 4.3.3 implies the following important corollary.

Corollary 4.3.4 *Let G be a noncyclic group of odd order n. Then*

$$\psi(G) < \frac{1}{2}\psi(C_n).$$

Notice that for groups of odd order Corollary 4.3.4 is stronger than our Theorem 4.3.1, which only claims that $\psi(G) \leq \frac{7}{11}\psi(C_n)$.

A significant ingredient of our proofs is the following result of H. Amiri, S.M. Jafarian Amiri, and I.M. Isaacs.

Theorem 4.3.5 (Corollary B in [1]) *If P is a cyclic normal Sylow p-subgroup of a finite group G, then*

$$\psi(G) \leq \psi(P)\psi(G/P),$$

with equality if and only if P is central in G.

Another important ingredient in our proofs is the following lower bound for the Euler's function $\varphi(n)$. We proved the following theorem.

Theorem 4.3.6 *Let n be an integer larger than 1, with the largest prime divisor p and the smallest prime divisor q. Then*

$$\varphi(n) \geq \frac{q-1}{p}n.$$

Proof Let

$$n = p_1^{r_1} p_2^{r_2} \cdots p_k^{r_k},$$

where the p_i's are primes, the r_i's are positive integers, and $p = p_1 > p_2 > \cdots > p_k = q$. Our proof is by induction on k.

If $k = 1$, then $n = p^{r_1}$ and

$$\varphi(n) = \varphi(p^{r_1}) = \frac{p-1}{p} p^{r_1} = \frac{p-1}{p} n,$$

as required.

Suppose now that $k > 1$ and that the theorem holds for all integers n with less than k distinct prime divisors. Set $m = p_2^{r_2} \cdots p_k^{r_k}$. Then $n = p_1^{r_1} m$, $\varphi(n) = \varphi(p_1^{r_1})\varphi(m)$ and since $p_2 \leq p_1 - 1$, it follows by induction that

$$\varphi(m) \geq \frac{p_k - 1}{p_2} m \geq \frac{p_k - 1}{p_1 - 1} m.$$

As $\varphi(p_1^{r_1}) = \frac{p_1 - 1}{p_1} p_1^{r_1}$, it follows that

$$\varphi(n) \geq \frac{p_1 - 1}{p_1} p_1^{r_1} \frac{p_k - 1}{p_1 - 1} m = \frac{p_k - 1}{p_1} n$$

or equivalently $\varphi(n) \geq \frac{q-1}{p} n$, as required. The proof is now complete. □

In the following subsection, we shall sketch the proof of Theorem 4.3.3. This proof is simpler than that of Theorem 4.3.1, but it includes the main ingredients of the proof of Theorem 4.3.1.

4.4 A Sketch of the Proof of Theorem 4.3.3

Theorem 4.3.3 *Let G be a noncyclic group of order n and let q be the smallest prime divisor of n. Then*

$$\psi(G) < \frac{1}{q-1} \psi(C_n).$$

Proof We need to prove that if G is a **noncyclic** group of order n, then $\psi(G) < \frac{1}{q-1} \psi(C_n)$, where q denotes the smallest prime divisor of n.

In other words, we need to prove that if G is a group of order n satisfying

$$\psi(G) \geq \frac{1}{q-1}\psi(C_n),$$

then $G = C_n$.

Recall that $\varphi(n)$ is the number of elements of C_n of order n. Hence, $\psi(C_n) > n\varphi(n)$ and by Theorem 4.3.6 $\varphi(n) \geq (q-1)n/p$, where p denotes the largest prime divisor of n. Thus, by our assumptions

$$\psi(G) \geq \frac{1}{q-1}\psi(C_n) > \frac{n}{q-1}\varphi(n) \geq \frac{n^2}{p}.$$

Hence, $\psi(G) > \frac{n^2}{p}$, which implies that there exists $x \in G$ of order $o(x) > n/p$. Thus, $[G : \langle x\rangle] < p$, so $\langle x\rangle$ contains a cyclic Sylow p-subgroup P of G.

Since P is normal in $\langle x\rangle$, it follows that $N_G(P) \geq \langle x\rangle$, which implies that $[G : N_G(P)] < p$. Hence, $N_G(P) = G$ and P is a cyclic normal Sylow p-subgroup of G.

Thus, the assumptions of Theorem 4.3.5 are satisfied and it follows that $\psi(G) \leq \psi(P)\psi(G/P)$. On the other hand, since $n = |G| = |P|\frac{n}{|P|}$ and $(|P|, \frac{n}{|P|}) = 1$, it follows from our assumptions that

$$\psi(G) \geq \frac{1}{q-1}\psi(C_n) = \frac{1}{q-1}\psi(C_{|P|})\psi(C_{\frac{n}{|P|}}).$$

The two inequalities concerning $\psi(G)$ imply that

$$\psi(P)\psi(G/P) \geq \psi(G) \geq \frac{1}{q-1}\psi(C_{|P|})\psi(C_{\frac{n}{|P|}}),$$

and cancelation by $\psi(P) = \psi(C_{|P|})$ yields

$$\psi(G/P) \geq \frac{1}{q-1}\psi(C_{\frac{n}{|P|}}).$$

If $n = p^r$, p a prime, then the existence of $x \in G$ satisfying $o(x) > n/p$ implies that $o(x) = n$ and G is cyclic, as required. So we may assume that $n = |G|$ is divisible by exactly k different primes with $k > 1$. Applying induction with respect to k, we may assume that the theorem holds for groups of order which has less than k distinct prime divisors.

Now $|G/P|$ has $k-1$ distinct prime divisors, $|G/P| = \frac{n}{|P|}$ and as shown above, G/P satisfies $\psi(G/P) \geq \frac{1}{q-1}\psi(C_{\frac{n}{|P|}})$, which is our assumption. Therefore, by our inductive hypothesis, G/P is cyclic. Denoting by F the cyclic complement of P in G, we deduce that

$$G = P \rtimes F.$$

Since P and F are cyclic, $|P||F| = |G| = n$ and $(|P|, |F|) = 1$, it follows that $\psi(C_n) = \psi(P)\psi(F)$.

The rest of the proof is technical. Recall that $G = P \rtimes F$ and $(|P|, |F|) = 1$. If $C_F(P) = F$, then $G = P \times F$ and G is cyclic, as required.

So it suffices to prove that if $C_F(P) = Z < F$, then $\psi(G) < (1/(q-1))\psi(C_n)$, contrary to our assumptions. We showed that if $C_F(P) = Z < F$, then

$$\psi(G) = \psi(P)\psi(Z) + |P|\psi(F \setminus Z),$$

which implies that $\psi(G) < \psi(P)\psi(Z) + |P|\psi(F)$. Thus

$$\psi(G) < \psi(P)\psi(F)\Big(\frac{\psi(Z)}{\psi(F)} + \frac{|P|}{\psi(P)}\Big)$$

and since $\psi(C_n) = \psi(P)\psi(F)$, it follows that

$$\psi(G) < \psi(C_n)\Big(\frac{\psi(Z)}{\psi(F)} + \frac{|P|}{\psi(P)}\Big).$$

We concluded the proof by showing that if $C_F(P) = Z < F$, then

$$\frac{\psi(Z)}{\psi(F)} + \frac{|P|}{\psi(P)} < \frac{1}{q-1}.$$

Hence, $\psi(G) < \psi(C_n)\frac{1}{q-1}$, a final contradiction. The proof is now complete. □

4.5 *Other Results of Herzog, Longobardi, and Maj*

Let G be a noncyclic group of order n and let q be the least prime divisor of n. As stated before, our main results were

Theorem 4.3.1. $\psi(G) \le \frac{7}{11}\psi(C_n)$.

Theorem 4.3.3. $\psi(G) < \frac{1}{q-1}\psi(C_n)$.

But what can we say about groups of order n satisfying

$$\psi(G) \ge \frac{1}{q}\psi(C_n)?$$

There exist noncyclic groups satisfying this condition. For example, $\psi(S_3) = 13 > \frac{1}{2}\psi(C_6) = \frac{21}{2}$.

We tackled a more general problem: which groups satisfy $\psi(G) \ge \frac{1}{2(q-1)}\psi(C_n)$? Our result was the following theorem:

Theorem 4.5.1 *Let $|G| = n$ and let q and p be the smallest and the largest prime divisors of n, respectively. Suppose that G satisfies*

$$\psi(G) \geq \frac{1}{2(q-1)}\psi(C_n).$$

Then, G is soluble and either its Sylow p-subgroups or its Sylow q-subgroups are cyclic.

Theorem 4.5.1 implies the following corollary:

Corollary 4.5.2 *The conclusions of Theorem 4.5.1 hold under each of the following conditions:*

(a) $|G| = n$ and $\psi(G) \geq \frac{1}{q}\psi(C_n)$.
(b) $|G| = n$ is odd and $\psi(G) \geq \frac{1}{q+1}\psi(C_n)$.

Proof (a) Since $q \geq 2$, it follows that $2(q-1) \geq q$ and hence

$$\psi(G) \geq \frac{1}{q}\psi(C_n) \geq \frac{1}{2(q-1)}\psi(C_n).$$

(b) Since $q \geq 3$, it follows that $2(q-1) \geq q+1$ and hence

$$\psi(G) \geq \frac{1}{q+1}\psi(C_n) \geq \frac{1}{2(q-1)}\psi(C_n).$$

□

Notice that the condition $\psi(G) \geq \frac{1}{2(q-1)}\psi(C_n)$ of Theorem 4.5.1 implies that

$$\psi(G) \geq \frac{1}{2(q-1)}\psi(C_n) > \frac{1}{2(q-1)}n\varphi(n)$$

and by Theorem 4.3.6 $\psi(G) > \frac{n}{2(q-1)}\frac{q-1}{p}n = \frac{n^2}{2p}$. Hence, there exists $x \in G$ satisfying $o(x) > \frac{n}{2p}$ and consequently $[G : \langle x\rangle] < 2p$.

Concerning groups satisfying $[G : \langle x\rangle] < 2p$, we obtained the following result:

Proposition 4.5.3 *If there exists $x \in G$ such that*

$$[G : \langle x\rangle] < 2p,$$

where p is the maximal prime divisor of $|G|$, then one of the following statements holds:

(i) G has a normal cyclic Sylow p-subgroup,
(ii) G is soluble and $\langle x\rangle$ is a maximal subgroup of G of index either p or $p+1$.

Finally, we mention our $\psi(G)$-based sufficient condition for solubility of finite groups.

Theorem 4.5.4 *Let $|G| = n$ and suppose that G satisfies the following inequality:*

$$\psi(G) \geq \frac{3}{5}n\varphi(n).$$

Then, G is soluble and $G'' \leq Z(G)$.

This condition is certainly not necessary for the solubility of G. For example, for $n = 8$, we have $\psi(C_2 \times C_2 \times C_2) = 15 < \frac{3}{5} \cdot 8 \cdot 4 = \frac{3}{5}n\varphi(n)$.

On the other hand, for $n = 60$, the simple group A_5 satisfies $\psi(A_5) = 211 > \frac{1}{5}n\phi(n) = 192$. Thus, the statement "If $|G| = n$ and $\psi(G) \geq \frac{1}{5}n\varphi(n)$, then G is soluble" is incorrect.

The reader can consult papers [19, 20] for some recent results.

For the proof of Theorem 4.5.4, we used the following identity of Ramanujan (see [32], p. 46):

Theorem 4.5.5 (Ramanujan) *If*

$$q_1 = 2, q_2, \cdots, q_n, \cdots$$

is the increasing sequence of all primes, then

$$\prod_{i=1,\cdots,\infty} \frac{q_i^2+1}{q_i^2-1} = \frac{5}{2},$$

What we needed was the following lemma:

Lemma 4.5.6 *Let $p_2, p_3, \ldots, p_s$ be primes satisfying $p_2 < p_3 < \cdots < p_s$. If $p_2 > 3$, then*

$$\prod_{i=2}^{s} \frac{p_i^2-1}{p_i^2+1} > \frac{5}{6}.$$

Proof Since $p_2 > 3$, Theorem 4.5.5 implies that

$$\frac{2^2+1}{2^2-1}\frac{3^2+1}{3^2-1}\prod_{i=2}^{s}\frac{p_i^2+1}{p_i^2-1} = \frac{5}{3}\frac{10}{8}\prod_{i=2}^{s}\frac{p_i^2+1}{p_i^2-1} < \frac{5}{2}.$$

Thus

$$\prod_{i=2}^{s}\frac{p_i^2+1}{p_i^2-1} < \frac{6}{5},$$

yielding

$$\prod_{i=2}^{s} \frac{p_i^2 - 1}{p_i^2 + 1} > \frac{5}{6},$$

as required. □

4.6 The Minimal Value of $\psi(G)$

As we have seen, the **maximal** value of $\psi(G)$ for groups of order n is attained by the unique group C_n.

The **minimal** value of $\psi(G)$ for groups of order n had been also investigated. For p-groups of order p^r, the minimal value of $\psi(G)$ is attained by groups of exponent p. Thus, for 2-groups of order 2^r, the group attaining the minimal value is unique, the elementary abelian group of order 2^r. But for $p > 2$, groups of order $p^r > p^2$ and of exponent p are not uniquely determined. Therefore, in these cases, groups with the minimal value of $\psi(G)$ are not uniquely determined. For example, the two groups of order 3^3 and exponent 3 are both groups with minimal value of $\psi(G)$ among groups of order 27.

What can we say in the general case of groups of order n? Very little is known in that direction. The following result was obtained in 2011 by H. Amiri and S. M. Jafarian Amiri in [2]:

Theorem 4.6.1 *Let G be a nilpotent group of order n and suppose that there are non-nilpotent groups of order n. Then, there exists a non-nilpotent group K of order n satisfying $\psi(K) < \psi(G)$.*

They also proved the following proposition.

Proposition 4.6.2 *Among all nilpotent groups of order n, the groups with the minimal value of $\psi(G)$ are those with all Sylow subgroups of prime exponent.*

The problem of the minimal value of $\psi(G)$ for groups of order n was also investigated in relation to the non-abelian simple groups. For the six simple groups of the smallest orders:

$$|A_5| = 60, \ |PSL(3,2)| = 168, \ |A_6| = 360, \ |PSL(2,8)| = 504,$$
$$|PSL(2,11)| = 660 \quad \text{and} \quad |PSL(2,13)| = 1092,$$

the values of $\psi(G)$ are

$$\psi(A_5) = 211, \ \psi(PSL(3,2)) = 715, \ \psi(A_6) = 1411, \ \psi(PSL(2,8)) = 3319,$$
$$\psi(PSL(2,11)) = 3741 \quad \text{and} \quad \psi(PSL(2,13)) = 7281.$$

In all these cases, the $\psi(G)$ of these simple groups is the unique minimum of the values of $\psi(G)$ for groups of the corresponding orders. This observation raised several questions.

Question 4.6.3 *If S is a simple group of order n, is $\psi(S)$ the minimal value of $\psi(G)$ for groups of order n?*

The answer is "NO". For example, there are two simple groups of order 20160: A_8 and $PSL(3, 4)$. Now $\psi(A_8) = 137047$, while $\psi(PSL(3, 4)) = 103111$. Hence, $\psi(A_8)$ is not minimal. We do not know if the value $\psi(PSL(3, 4)) = 103111$ is minimal.

Question 4.6.4 *If S is a finite simple group and G is a non-simple group satisfying $|G| = |S|$, does it follow that $\psi(S) < \psi(G)$?*

Here also the answer is "NO". It was shown by Y. Marefat, A. Iranmanesh and A. Tehranian in 2013 (see [42]) that if $S = PSL(2, 64)$ and $G = 3^2 \times Sz(8)$, then $|G| = |S|$ and $\psi(G) \leq \psi(S)$.

We still conjecture that the following is true:

Conjecture 4.6.5 *If S is a simple group and G is a soluble group satisfying $|G| = |S|$, then $\psi(S) < \psi(G)$.*

5 Results Concerning Some Other Functions

5.1 The Theorems of Ladisch and of Jafarian Amiri and M. Amiri

Definitions 5.1.1 (a) Let A and B be finite sets such that $|A| = |B|$ and for each $a \in A$ and $b \in B$ the orders of a and b, denoted by $o(a)$ and $o(b)$, are defined. A bijection $\sigma : A \to B$ is called an *order multiplying bijection*, or an *OM*-bijection in short, if for each $a \in A$, $\sigma(a)$ satisfies

$$o(a) \mid o(\sigma(a)).$$

The following conjecture is still open, and it is Question 18.1 in the Kourovka Notebook (see [31]):

Conjecture 5.1.2 *If G is a finite group of order n, then there exists an OM-bijection*

$$\sigma : G \to C_n.$$

We shall now present in detail the Frieder Ladisch's proof of the conjecture for soluble groups (see [6]).

Theorem 5.1.3 *Let G be a finite soluble group of order n. Then, there exists an OM-bijection*

$$\sigma : G \to C_n.$$

Proof Let G be a minimal counterexample. Then, $|G| > 1$ and there exists a normal elementary abelian p-subgroup N of G. By the minimality of G, there exists an OM-bijection

$$\sigma : G/N \to C_{|G/N|},$$

and it suffices to prove that σ can be lifted to an OM-bijection between $G \to C_n$.

Since $C_{|G/N|}$ is isomorphic to $C_{|G|}/C_{|N|}$, we may assume that σ is an OM-bijection

$$\sigma : G/N \to C_{|G|}/C_{|N|}.$$

We shall use overbars to denote the canonical epimorphisms $G \to G/N$ and $C_{|G|} \to C_{|G|}/C_{|N|}$. Since σ is an OM-bijection, for each $g \in G$ and $c \in C_{|G|}$ such that $\bar{c} = \sigma(\bar{g})$ we have

$$o(\bar{g}) \mid o(\bar{c}). \tag{1}$$

Thus, it suffices to find for every $g \in G$ an OM-bijection between the cosets

$$gN \to cC_{|N|},$$

where $c \in C_{|G|}$ and $\bar{c} = \sigma(\bar{g})$.

For our proof, we need the following three preliminary results:

(a) If $g \in G$, then either $o(g) = o(\bar{g})$ or $o(g) = po(\bar{g})$.

(b) If $p \nmid |\bar{c}|$, then the orders of elements in $cC_{|N|}$ are $o(\bar{c})p^k$, where $p^k \mid |N|$ and $o(\bar{c})$ occurs exactly once.

(c) If $p \mid o(\bar{c})$, then all elements in $cC_{|N|}$ have order $o(\bar{c})|N|$.

First, we prove (a). Since $o(\bar{g})$ is the minimal positive integer such that $g^{o(\bar{g})} \in N$ and N is an elementary abelian p-group, it follows that either $o(g) = o(\bar{g})$ or $o(g) = po(\bar{g})$.

Now we prove (b). Let $o(\bar{c}) = m$ and $p \nmid m$. Then, $c^m \in C_{|N|}$ and for each $x \in C_{|N|}$, m is the minimal positive integer such that $(cx)^m \in C_{|N|}$. Hence, $m \mid o(cx)$ and

$$o(cx) = mo((cx)^m) = o(\bar{c})p^k$$

for some $p^k \mid |N|$. Since $(cx)^m = c^m x^m$ and $x \in C_{|N|}$, our assumption that $(m, p) = 1$ implies that there is a unique $x \in C_{|N|}$ satisfying $(cx)^m = 1$. Hence, the orders of elements in $cC_{|N|}$ are as claimed.

Finally, we prove (c). Let $o(\bar{c}) = pm$ and let $w \in cC_{|N|}$. Then, $o(\bar{w}) = o(\bar{c}) = pm$ and $w^m \in C_{|G|} \setminus C_{|N|}$. Since $(w^m)^p \in C_{|N|}$, it follows that $|\langle w^m \rangle C_{|N|}| = p|N|$ and $o(w^m) = p|N|$. Since pm is minimal such that $w^{pm} \in C_{|N|}$, it follows that m is minimal such that $w^m \in C_{p|N|}$, and hence $o(w) = mo(w^m) = mp|N| = o(\bar{c})|N|$, as claimed.

We return now to the proof of the theorem. The proof will be divided into two cases:

Case 1: $p \nmid o(\bar{c})$.

Case 2: $p \mid o(\bar{c})$.

We begin with Case 1. Since $p \nmid o(\bar{c})$ and by (1) $o(\bar{g}) \mid o(\bar{c})$, it follows that $p \nmid o(\bar{g})$. Let $o(\bar{g}) = m$, where $p \nmid m$. Then, $g^m \in N$, say $g^m = y \in N$ and $gy = yg$.

Since y is a p-element and $p \nmid m$, there exists a unique element $u \in \langle y \rangle$ such that $u^m = y^{-1}$. For this $u \in N$, we have $1 = g^m y^{-1} = g^m u^m = (gu)^m$ and $gu \in gN$ satisfies $p \nmid o(gu)$. We map gu to the unique element of $cC_{|N|}$ of order $o(\bar{c})$ (see (b)). It follows from (a) that $o(gu) = o(\bar{g})$, and as by (1) $o(\bar{g}) \mid o(\bar{c})$, we may conclude that $o(gu) \mid o(\bar{c})$, as required.

The rest of gN we map arbitrarily onto the rest of $cC_{|N|}$, which by (b) consists of element of order $o(\bar{c})p^k > o(\bar{c})$. By applying (a) and (1) to each such $gw \in gN$, we get

$$o(gw) \mid po(\bar{g}) \mid po(\bar{c}) \mid o(\bar{c})p^k,$$

as required. Therefore, this mapping is an OM-bijection.

Finally, we approach Case 2. Since $p \mid o(\bar{c})$, it follows by (c) that all elements in $cC_{|N|}$ have order $o(\bar{c})|N|$. Let $h \in gN$. Then, (a) and (1) again imply that

$$o(h) \mid po(\bar{h}) \mid po(\bar{c}) \mid |N|o(\bar{c}).$$

Hence, any mapping of gN onto $cC_{|N|}$ is an OM-bijection, as required.

The proof of the theorem is complete. □

In addition to the abovementioned Conjecture 5.1.2, it is worthwhile to consider the following related questions:

Question 5.1.4 *Does the Conjecture 5.1.2 hold for finite simple groups?*

Question 5.1.5 *If Conjecture 5.1.2 holds for finite simple and soluble groups, does it imply that it holds for all finite groups?*

OM-bijections were also considered in the recent paper [30] of S. M. Jafarian Amiri and M. Amiri. They introduced the following definition.

Definition 5.1.6 Let H be a finite group. Then, $[H]$ denotes the set of all finite groups G for which there is an OM-bijection $f : G \to H$.

Remark 5.1.7 If Conjecture 5.1.2 holds, then $[C_n]$ consists of all groups of order n. By Theorem 5.1.3, $[C_n]$ certainly contains all soluble groups of order n.

Remark 5.1.8 Let π be a set of primes and let G_π denote the set of π-elements in the finite group G. If $G \in [H]$, then $|H_\pi| \leq |G_\pi|$, since if $f : G \to H$ is an OM-bijection, then $f^{-1}(H_\pi) \subseteq G_\pi$. In particular, since $|G| = |H|$, it follows that if the Sylow p-subgroup of G is normal in G, then so is the Sylow p-subgroup of H.

The main result of [30] is the following theorem.

Theorem 5.1.9 *Let G be a soluble group of order n which is divisible by the prime p and suppose that a Sylow p-subgroup P of G is neither cyclic nor generalized quaternion. Then, $G \in [C_{\frac{n}{p}} \times C_p]$.*

Concerning Theorem 5.1.9, notice that if G has a cyclic Sylow p-subgroup of order $p^a > p$, then $G \notin [C_{\frac{n}{p}} \times C_p]$ since G contains an element of order p^a, but $C_{\frac{n}{p}} \times C_p$ does not contain such an element. Moreover, if Q_{2^n} is the generalized quaternion group, then $Q_{2^n} \notin [C_{2^{n-1}} \times C_2]$ since Q_{2^n} has only one involution, but $C_{2^{n-1}} \times C_2$ has three involutions.

The authors of [30] raised the following questions:

Question 5.1.10 *Let G be a non-soluble group of order n which is divisible by the prime p and suppose that a Sylow p-subgroup of G is neither cyclic nor generalized quaternion. Is it true that $G \in [C_{\frac{n}{p}} \times C_p]$?*

Question 5.1.11 *Let S be a finite simple group and let $G \in [S]$. Is G necessarily simple?*

They also proved the following related proposition.

Proposition 5.1.12 *Suppose that G is a finite group satisfying $G \in [H]$. Then, the following statements hold.*

(1) If G is nilpotent, then also H is nilpotent.
(2) If G is p-nilpotent, then also H is p-nilpotent.

Notice that (1) follows immediately by Remark 5.1.8. The proof of (2) is a bit more complicated.

5.2 The Results of Garonzi and Patassini

In their paper [10], M. Garonzi and M. Patassini considered other problems related to the orders of elements of finite groups G. In particular, they dealt with the function

$$P(G) := \prod_{x \in G} o(x)$$

and with the functions

$$R_G(r, s) := \sum_{x \in G} \frac{o(x)^s}{\varphi(o(x))^r},$$

where r, s are real numbers and φ denotes the Euler's totient function. If $r = s$, then the function $R_G(r, r)$ is denoted by $R_G(r)$.

Notice that

$$R_G(0, s) = \sum_{x \in G} o(x)^s.$$

In particular, $R_G(0, 1) = \sum_{x \in G} o(x) = \psi(G)$.

Recall the statement of Theorem 4.2.1 of H. Amiri, S. M. Jafrian Amiri and I. M. Isaacs:

Theorem 4.2.1 *If $|G| = n$, then $\psi(G) \leq \psi(C_n)$, with equality if and only if G is cyclic.*

In [10], this result is generalized in two directions.

First, we mention Theorem 5(3) of [10]:

Theorem 5.2.1 *If $|G| = n$, $r \leq s - 1$ and $s \geq 1$, then $R_G(r, s) \leq R_{C_n}(r, s)$, with equality if and only if G is cyclic.*

This theorem implies the following.

Corollary 5.2.2 *If $|G| = n$, $s \geq 1$ and $r = 0$, then*

$$\sum_{x \in G} o(x)^s = R_G(0, s) \leq R_{C_n}(0, s) = \sum_{x \in C_n} o(x)^s,$$

with equality if and only if G is cyclic.

The case $s = 1$ corresponds to Theorem 4.2.1.

Next, we mention a weaker version of Theorem 5(1) in [10]:

Theorem 5.2.3 *If $|G| = n$, $s < r$ and $s \leq 0$, then $R_G(r, s) \geq R_{C_n}(r, s)$, with equality if and only if G is cyclic.*

This theorem implies the following.

Corollary 5.2.4 *If $|G| = n$, $r = 0$ and $s < 0$, then*

$$\sum_{x \in G} o(x)^s = R_G(0, s) \geq R_{C_n}(0, s) = \sum_{x \in C_n} o(x)^s,$$

with equality if and only if G is cyclic.

Thus, for negative powers of orders of group elements, the direction of the inequality is opposite to that which occurs for powers greater or equal to 1.

Notice that Conjecture 5.1.2 would immediately imply the inequalities in Corollaries 5.2.2 and 5.2.4. By Laudisch's Theorem 5.1.3, this is the case if G is a soluble group.

The authors also proved the following result, which is similar to Theorem 4.2.1.

Theorem 5.2.5 *If $|G| = n$ then $P(G) \leq P(C_n)$, with equality if and only if G is cyclic.*

Notice also that if G is an arbitrary finite group and $x \in G$, then the cyclic group $\langle x \rangle$ has $\varphi(o(x))$ generators. This observation implies that

$$R_G(1) = \sum_{x \in G} \frac{o(x)}{\varphi(o(x))}$$

is equal to the sum of the sizes of the cyclic subgroups of G and

$$R_G(1, 0) = \sum_{x \in G} \frac{1}{\varphi(o(x))}$$

is equal to the number of cyclic subgroups of G. In that direction, the authors proved the following interesting theorem. Let $d(n)$ denote the number of positive divisors of the integer n.

Theorem 5.2.6 *Let G be a finite group. Then, G has at least $d(|G|)$ cyclic subgroups and G has exactly $d(|G|)$ cyclic subgroups if and only if G is cyclic.*

The authors also constructed for each positive integer γ infinitely many finite groups G with exactly $d(|G|) + \gamma$ cyclic subgroups.

Finally, we mention a very interesting characterization of nilpotency which was proved in [10].

Theorem 5.2.7 *If $|G| = n$ and $r < 0$ is a real number, then*

$$\sum_{x \in G} \left(\frac{o(x)}{\varphi(o(x))} \right)^r = R_G(r) \geq R_{C_n}(r) = \sum_{x \in C_n} \left(\frac{o(x)}{\varphi(o(x))} \right)^r,$$

with equality if and only if G is nilpotent.

Thus, we may detect the nilpotency of a finite group by looking at the functions $R_G(r, s)$ on the orders of all its elements. It follows by Corollary 5.2.4 that also the cyclicity can be detected in a similar way. The following question, related to the question 3.2.4, was raised by John Thompson:

Question 5.2.8 *Can the solubility of a finite group be detected by looking at the functions $R_G(r, s)$ on the orders of all its elements?*

As far as we know, this question is still open.

Garonzi and Patassini also proved the following result concerning noncyclic nilpotent groups (see Theorem 5.4 in [10]).

Proposition 5.2.9 *If G is a noncyclic nilpotent group of order n, then the following statements hold.*

(1) If $r > s$, then $R_G(r, s) > R_{C_n}(r, s)$.
(2) If $r < s$, then $R_G(r, s) < R_{C_n}(r, s)$.

We conclude this section with the following two questions raised in [10].

Question 5.2.10 *Let $|G| = n$. Does the equation*

$$\sum_{x\in G} \frac{o(x)^s}{\varphi(o(x))^r} = \sum_{x\in C_n} \frac{o(x)^s}{\varphi(o(x))^r} \tag{2}$$

detect solubility of G for some (r, s)? (This is a stronger version of Question 5.2.8).

Question 5.2.11 *Let $|G| = n$. What other structural properties of G can be detected by the equation (2) (for fixed (r, s))?*

The proofs of the abovementioned results are quite intricate, and therefore we refer the interested readers to the original paper.

5.3 The Results of De Medts and Tarnauceanu

In [7], T. De Medts and M. Tarnauceanu proved the following theorem.

Theorem 5.3.1 *If G is a nilpotent group of order n, then the multisets $\{\frac{o(x)}{\varphi(o(x))} : x \in G\}$ and $\{\frac{o(x)}{\varphi(o(x))} : x \in C_n\}$ are identical.*

Indeed, if P is a Sylow p-subgroup of G and Q is a Sylow p-subgroup of C_n, then the corresponding multisets $\{\frac{o(x)}{\varphi(o(x))} : x \in P\}$ and $\{\frac{o(x)}{\varphi(o(x))} : x \in Q\}$ are both equal to the multiset $\{1, (|P| - 1) \cdot \frac{p}{p-1}\}$. Since the functions o and φ are multiplicative, the statement of the theorem follows.

This observation yields the following corollary.

Corollary 5.3.2 *If G is a nilpotent group of order n and r is a real number, then*

$$\sum_{x\in G} \left(\frac{o(x)}{\varphi(o(x))}\right)^r = \sum_{x\in C_n} \left(\frac{o(x)}{\varphi(o(x))}\right)^r.$$

In particular

$$\sum_{x\in G} \frac{o(x)}{\varphi(o(x))} = \sum_{x\in C_n} \frac{o(x)}{\varphi(o(x))}.$$

We conclude this survey with a couple of conjectures of De Medts and Tarnauceanu. In [7], they conjectured that if $R_G(1) = R_{C_n}(1)$, then G is nilpotent:

Conjecture 5.3.3 *If* $|G| = n$ *and*

$$\sum_{x \in G} \frac{o(x)}{\varphi(o(x))} = \sum_{x \in C_n} \frac{o(x)}{\varphi(o(x))},$$

then G is nilpotent.

Moreover, in [8], they conjectured that $R_G(1) \leq R_{C_n}(1)$ for all groups of order n:

Conjecture 5.3.4 *If* $|G| = n$, *then*

$$\sum_{x \in G} \frac{o(x)}{\varphi(o(x))} \leq \sum_{x \in C_n} \frac{o(x)}{\varphi(o(x))}.$$

References

1. H. Amiri, S.M. Jafarian Amiri, I.M. Isaacs, Sums of element orders in finite groups. Commun. Algebra **37**, 2978–2980 (2009)
2. H. Amiri, S.M. Jafarian Amiri, Sum of element orders on finite groups of the same order. J. Algebra Appl. **10**(2), 187–190 (2011)
3. R. Brandl, W.J. Shi, Finite groups whose elements are consecutive integers. J. Algebra **143**(2), 388–400 (1991)
4. R. Brandl, W.J. Shi, The characterization of $PSL(2, q)$ by its element orders. J. Algebra **163**(1), 109–114 (1994)
5. A.A. Buturlakin, Sibirsk. Electron. Math. Rep. **7**, 111–114 (2010)
6. T. De Medts, *Order Increasing Bijection From Arbitrary Groups to Cyclic Groups* (2012). http://mathoverflow.net/questions/104183/
7. T. De Medts, M. Tarnauceanu, Finite groups determined by an inequality of the orders of their subgroups. Bull. Belg. Math. Soc. Simon Stevin **15**, 699–704 (2008)
8. T. De Medts, M. Tarnauceanu, *An Inequality Detecting Nilpotency of Finite Groups* (2012), http://arxiv.org/abs/1207.1020
9. G. Frobenius, Verallgemeinerung des Sylowschen Satzes (Berliner Sitz, 1895) pp. 981–993
10. M. Garonzi, M. Patassini, Inequalities detecting structural properties of a finite group. Commun. Algebra **45**(2), 677–687 (2017)
11. I.B. Gorshkov, Recognizability of alternating groups by spectrum. Algebra Log. **52**(1), 41–45 (2013)
12. W. Guo, A.S. Mamontov, On groups whose element orders divide 6 and 7 Sib. Math. J **58**(1), 67–71 (2017)
13. N.D. Gupta, V.D. Mazurov, On groups with small orders of elements. Bull. Austral. Math. Soc. **60**(2), 197–205 (1999)
14. M. Hall, Solution of the Burnside problem for exponent six. Ill. J. Math. **2**, 764–786 (1958)
15. M. Hall, *The Theory of Groups* (AMS Chelsea Publishing, Providence, 1999)
16. H. Heineken, F. Russo, Groups described by element numbers. Forum Math. **27**(4), 1961–1977 (2015)
17. H. Heineken, F. Russo, On a notion of breadth in the sense of Frobenius. J. Algebra **424**, 208–221 (2015)

18. M. Herzog, P. Longobardi, M. Maj, An exact upper bound for sums of element orders in non-cyclic finite groups. J. Pure Appl. Algebra **222**(7), 1628–1642 (2018)
19. M. Herzog, P. Longobardi, M. Maj, Sums of element orders in groups of order $2m$ with m odd. Commun. Algebra (to appear)
20. M. Herzog, P. Longobardi, M. Maj, Two new criteria for solvability of finite groups. J. Algebra **511**, 215–226 (2018)
21. N. Iiyori, H. Yamaki, On the conjecture of Frobenius. Bull. Am. Math. Soc. (N.S.) **25**(2), 413–416 (1991)
22. E. Jabara, On {2, 3}-groups without elements of order 6. Sib. Math. J. **57**(4), 744–746 (2016)
23. E. Jabara, Fixed point free action of groups of exponent 5. J. Aust. Math. Soc. **77**, 297–304 (2004)
24. E. Jabara, D.V. Lytkina, On groups of exponent 36 Sib. Math. J. **54**, 29–32 (2013)
25. E. Jabara, D.V. Lytkina, V.D. Mazurov, Some groups of exponent 72. J. Group Theory **17**, 947–955 (2014)
26. E. Jabara, D.V. Lytkina, A.S. Mamontov, Recognizing M_{10} by spectrum in the class of all groups. Int. J. Algebra Comput. **24**, 113–119 (2014)
27. E. Jabara, D.V. Lytkina, V.D. Mazurov, A.S. Mamontov, Groups whose element orders do not exceed 6. Algebra Log. **53**(5), 365–376 (2014)
28. E. Jabara, A.S. Mamontov, Recognition of the group $L_3(4)$ by the set of element orders in the class of all groups. Algebra Log. **54**(4), 279–282 (2015)
29. E. Jabara, A.S. Mamontov, On periodic groups with narrow spectrum. Sim. Math. J. **57**, 538–541 (2016)
30. S.M. Jafarian Amiri, M. Amiri, Characterization of finite groups by a bijection with a divisible property on the element orders. Commun. Algebra **45**(8), 3396–3401 (2017)
31. E.I. Khukhro, V.D. Mazurov, in *Unsolved Problems in Group Theory. The Kourowka Notebook. No. 18* (English version) (2017), arXiv:1401.0300v10
32. Le Lionnais, *Les nombres remarquables* (Hermann, Paris, 1983)
33. F. Levi, B.L. van der Warden, Über eine besondere Klasse von Gruppen. Abh. Math. Semin. Hambg. Univ. **9**, 154–158 (1933)
34. D.V. Lytkina, A.A. Kuznetsov, Recognizability by spectrum of the group $L_2(7)$ in the class of all groups. Sib. Elektron Mat. Izv **4**, 136–140 (2007)
35. D.V. Lytkina, V.D. Mazurov, {2, 3}-groups with no elements of order 6. Algebra Log. **55**(6), 1098–1104 (2014)
36. D.V. Lytkina, V.D. Mazurov, On groups of period 12 Sib. Math. J. **56**(3), 471–475 (2015)
37. D.V. Lytkina, V.D. Mazurov, A.S. Mamontov, The local finiteness of some groups of period 12. Sib. Math. J. **53**(6), 1005–1109 (2012)
38. D.V. Lytkina, V.D. Mazurov, A.S. Mamontov, E. Jabara, Groups whose element orders do not exceed 6. Algebra Log. **53**(5), 365–376 (2014)
39. D.V. Lytkina, The structure of a group with elements of order at most 4. Sib. Math. J. **48**(2), 283–287 (2007)
40. A.S. Mamontov, Groups of exponent 12 without elements of order 12. Sib. Math. J. **54**(1), 114–118 (2013)
41. A.S. Mamontov, V.D. Mazurov, On periodic groups with elements of small orders. Sib. Math. J. **50**(2), 316–321 (2009)
42. Y. Marefat, A. Iranmanesh, A. Tehranian, On the sum of element orders of finite simple groups. J. Algebra Appl. **12**(7), 1350026 (2013)
43. V.D. Mazurov, On the set of orders of elements of a finite group. Algebra Log. **33**(1), 49–55 (1994)
44. V.D. Mazurov, On groups of exponent 24. Algebra Log. **49**(6), 515–525 (2011)
45. V.D. Mazurov, Groups of period 60 with prescribed orders of elements. Algebra Log. **39**(3), 189–198 (2000)
46. V.D. Mazurov, W.J. Shi, On periodic groups with prescribed orders of elements. Sci. China Ser. A. Math. **52**(2), 311–317 (2009)

47. V.D. Mazurov, W.J. Shi, A criterion of unrecognizability by spectrum for finite groups. Algebra Log. **51**(2), 239–243 (2012)
48. V.D. Mazurov, A. Yu, A.I.Sozutov Ol'shanskii, Infinite groups of finite period. Algebra Log. **54**(2), 161–166 (2015)
49. W. Meng, Finite groups of global breath four in the sense of Frobenius. Commun. Algebra **45**(2), 660–665 (2017)
50. W. Meng, J. Shi, On an inverse problem to Frobenius' theorem. Arch. Math. (Basel) **96**(2), 109–114 (2011)
51. W. Meng, J. Shi, K. Chen, On an inverse problem to Frobenius' theorem II. J. Algebra Appl. **11**, 1250092 (2012). (8 pages)
52. B.H. Neumann, Groups whose elements have bounded orders. J. Lond. Math. Soc. **S1–12**(2), 195–198 (1937)
53. M.F. Newman, Groups of exponent dividing seventy. Math. Sci. **4**, 149–157 (1979)
54. P.S. Novikov, S.I. Adjan, Infinite periodic groups I, II, III Izv. Akad. Nauk SSSR Ser. Mat. **32**, 212–244, 251–524, 709–731 (1968)
55. C.E. Praeger, W.J. Shi, A characterization of some alternating and symmetric groups. Commun. Algebra **22**(5), 1507–1530 (1994)
56. D.J.S. Robinson, *A Course in the Theory of Groups* (Springer, New York, 1996)
57. I.N. Sanov, Solution of Burnside's problem for exponent 4 (Russian). Leningr. State Univ. Ann. [Uchenye Zapiski] Mat. Ser. **10**, 166–170 (1940)
58. R. Shen, G. Chen, Ch. Wu, On groups with the second largest value of the sum of element orders. Commun. Algebra **43**, 2618–2631 (2015)
59. W.J. Shi, The characterization of the sporadic simple groups by their element orders. Algebra Colloq. **1**(2), 159–166 (1994)
60. A.V. Vasil'ev, M.A. Grechkoseeva, V.D. Mazurov, Characterization of finite simple groups by spectrum and order. Algebra Log. **48**(6), 385–409 (2009)
61. H. Yamaki, A conjecture of Frobenius and the sporadic simple groups I. Commun. Algebra **11**, 2513–2518 (1983)
62. H. Yamaki, A conjecture of Frobenius and the simple groups of Lie type I. Arch. Math. **42**, 344–347 (1984)
63. H. Yamaki, A conjecture of Frobenius and the simple groups of Lie type II. J. Algebra **96**, 391–396 (1985)
64. H. Yamaki, A conjecture of Frobenius and the sporadic simple groups I. Math. Comput. **46**, 609–611 (1986)
65. E.I. Zelmanov, The solution of the restricted Burnside problem for groups of odd exponent. Izv. Akad. Nauk. SSSR **54**(1), 42–59 (1990)
66. E.I. Zelmanov, The solution of the restricted Burnside problem for 2-groups. Math. Sb. **182**(4), 568–569 (1991)

Calculating Subgroups with GAP

Alexander Hulpke

1 Introduction

One of the earliest questions posed for the development of group-theoretic algorithms have been the determination of the subgroups of a finite group G, as well as the associated lattice structure.

Since G acts on its subgroups, an obvious storage improvement is to store the subgroups as conjugacy classes, representing each class by a subgroup U and a transversal of coset representatives of $N_G(U)$ in G.

The purpose of this article is to survey the methods that are currently in use for such computations, not with an aim to supersede the original descriptions [12, 25, 27, 38] or to give an implementable description, but to give an overview of the methods employed. This should allow the reader to understand the interplay of the methods employed, computational tools required, scope of calculations, and potential for adaption or modifications by users.

On the way, we will indicate a number of open problems, whose solution would lead to improvements of theoretical or practical aspects of the algorithms.

While we shall point to the GAP functions that implement the respective functionality, we shall stop short of printing transcripts of system sessions, instead the reader is referred to the system documentation.

Neither is this paper intended as a complete survey of Computation Group theory over its history of at least 60 years. We thus do not aim to cite every relevant work but give preference to handbooks or summary articles that are often easier accessible.

We will illustrate the scope of calculations by assuming a contemporary (as of 2017) standard desktop machine with a 3.5 GHz processor (utilizing just a single core) and 8 GB of memory.

A. Hulpke (✉)
Department of Mathematics, Colorado State University, 1874 Campus Delivery,
Fort Collins, CO 80523-1874, USA
e-mail: hulpke@colostate.edu
URL: http://www.math.colostate.edu/~hulpke

N. S. N. Sastry and M. K. Yadav (eds.), *Group Theory and Computation*,
Indian Statistical Institute Series, https://doi.org/10.1007/978-981-13-2047-7_5

2 Tools Required

In general, we will represent a subgroup S of the finite group G by a set of generators, given as elements of G. One may think of G as the group containing all transformations of a given kind—for example, in the case of permutations a symmetric group S_n or even the finitary symmetric group on positive integers. Similarly, in the case of matrices, this group might be the full general linear group.

We thus need methods that allow us to determine for such a subgroup S:

- The order of S.
- Test whether an element of G is contained in S, and if so:
- Express an element of S as a word in the given generators of S, thus enabling us to evaluate homomorphisms.
- Write a presentation for S in a given generating set, thus testing whether a map on generators is a homomorphism. (In practice, one often does not use an arbitrary generating set, but a specific one that allows for a nicer presentation.)
- Determine a composition series, a chief series, and the radical $\mathrm{Rad}(G)$ (the largest solvable normal subgroup) of G, as well as a representation of $G/\mathrm{Rad}(G)$ as a permutation or matrix group.

For permutation groups, such functionality is obtained through a stabilizer chain data structure [24, Chap. 4], respectively [43]. For matrix groups, such functionality is provided by the data structure of a composition tree [2, 39]. These tools can be extended to groups of other classes of invertible transformations of a finite object using the "black-box" paradigm [6].

For solvable groups, polycyclic generating sets (that is a set of generators that is adapted to a composition series and allows for an effective normal form) provide such functionality [31], see also [24, Chap. 8].

2.1 Complexity

For solvable groups, polynomial-time algorithms are known for all of these tasks.

For permutation groups, the known algorithms are proven to be polynomial time, as long as no composition factor of type ${}^2G_2(q)$ occurs (in which case the result will still be correct, but the time bound is not known to hold.). In fact, the algorithms are almost linear (linear up to logarithmic factors) time in a Las Vegas probabilistic setting (see Sect. 2.2 below).

Open Problem 1 *Show that the groups ${}^2G_2(q)$ have a short presentation (formally defined in the sense of [5]). Such a result will allow the removal of the qualifier in the previous paragraph.*

The complexity situation for matrix groups [6] is as with permutation groups with one further complication: $\mathrm{GL}_n(q)$ contains cyclic subgroups (Singer cycles) of order

$q^n - 1$, and calculations in these groups are equivalent to Discrete Logarithm problems. The proven complexity is therefore also up to a Discrete Logarithm "oracle", that is the cost of discrete logarithm calculations is not accounted for.

These polynomial-time algorithms for solvable and for permutation groups have been fully implemented in GAP and in Magma. The available implementations for the matrix group algorithms involve many, but not all, of the polynomial-time methods. The reason for this is that there are number of algorithms for subtasks that perform better in practice than the generic black-box algorithm, but so far no proof of polynomial time has been found.

Arbitrary finitely presented groups will require the use of a faithful representation in one the representations discussed before.

2.2 *Random Elements*

Some of the algorithms utilize random selections of elements. It thus seems appropriate to briefly address this issue.

First, on the computer random selection is always based on a random number generator, and thus is inherently pseudo-random.

Second, once we can test membership of elements, the underlying data structures allow us to construct a bijection between G and the numbers $1, \ldots, |G|$ and thus select elements of the same random quality as the random number generator provides.

Some of the functions to build basic data structures also utilize pseudo-random elements which are obtained as pseudo-random products of generators and inverses [4, 15]. All of these calculations then involve verification steps that ensure the returned result is always correct, regardless of the random choices or the quality of randomness.

As far as complexity is concerned, any such algorithm then lies in a class denoted by "Las Vegas": That is the algorithm will always return a correct result and will, with a user-chosen probability $0 < \varepsilon < 1$, terminate in the given time. However, with probability $1 - \varepsilon$, the calculation will take longer (but will eventually terminate with a correct result).

2.3 *Mid-level Tools*

Building on these tools, a number of mid-level tools obtain structural group-theoretic information

- For $S \leq G$, representatives of the cosets of S in G [17].
- The centralizer $C_G(g)$ of elements $g \in G$ as well as conjugating elements x that for a given $g, h \in G$ satisfy $g^x = h$ (if they exist). (For permutation groups, this is a backtrack search, following [33]).

- The normalizer $N_G(S)$ of a subgroup $S \leq G$ as well as conjugating elements x that for a given $S, T \leq G$ satisfy that $S^x = T$ [32].
- Representatives of the conjugacy classes of elements of G [8, 11, 26, 28, 37].
- Representatives of Sylow subgroups of G for a chosen prime.
- For a normal subgroup $N \lhd G$, representatives of the G-classes of complements to N in G, provided that N is solvable [14]. This algorithm is based on cohomology through a presentation for G/N.
  If G/N is solvable, complements can be computed in a combination of cohomology and reduction to subgroups [27].
- Determine an effective[1] isomorphism between two groups G and H (or show that no such isomorphism can exist) [9, 40].

These algorithms are typically not of polynomial, but exponential worst-case time complexity. However, in most cases of practical interest, they tend to work well, allowing for them to be used as building blocks for larger calculations.

3 The Basic Structure

The basic structure underlying most subgroup calculations and the one we shall use is based on the *solvable radical* (or *trivial fitting*) paradigm [3, 8, 22], as depicted in Fig. 1:

Let G be a finite group, $R = \mathrm{Rad}(G)$ and $\varphi\colon G \to G/R =: F$. Then, $S = \mathrm{Soc}(F) = \prod T_i$ must be the direct product of nonabelian simple groups T_i. We thus can assume that F is represented as a subgroup of $\mathrm{Aut}(\mathrm{Soc}(F))$; that is as a subgroup of a direct product of groups of the form $\mathrm{Aut}(T_i) \wr S_{m_i}$ for T_i simple and $\sum_i m_i$ the number of simple factors of S.

The action of F on the socle factors has a kernel denoted by $Pker$, the factor $Pker/S$ is a direct product of subgroups of outer automorphisms. We denote by $\underline{S}$ and $\underline{Pker}$ the full preimages of these subgroups in G.

We now determine subgroups in the following way.

1. Subgroups of the simple socle factors T_i.
2. Combine these to subgroups of $\mathrm{Soc}(F)$.
3. Calculate the subgroups of $F/\mathrm{Soc}(F)$ (which will be a much smaller group than F).
4. Extend the subgroups of $\mathrm{Soc}(F)$ to subgroups of F using the subgroups of $F/\mathrm{Soc}(F)$.
5. Determine a series of normal subgroups $R = R_0 > R_1 > R_2 > \cdots > R_k = \langle 1 \rangle$ with $R_i \lhd G$ and R_i/R_{i+1} elementary abelian.
6. Determine subgroups of G/R_{i+1} from subgroups of G/R_i (initialized for $i = 0$ with $G/R_0 = F$) and the G-module action on R_i/R_{i+1}. Iterate.

---

[1] Meaning that it, and its inverse can be applied to group elements to obtain the image.

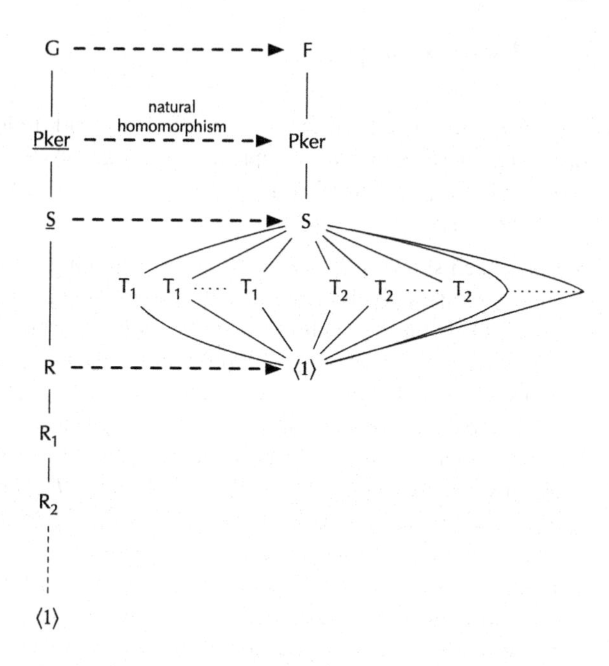

Fig. 1 Subgroups related to the solvable radical data structure

Typically, we will store not all subgroups of a group G, but only representatives of the conjugacy classes under G, since this saves substantially on the memory requirements. This enumeration up to conjugacy can be translated for each of these steps to conjugacy under suitable actions. For example, in step 1, it is conjugacy by the subgroup of $\mathrm{Aut}(T_i)$ induced through the action of $N_F(T_i)$. Finding representatives up to conjugacy can, in general, mean that we have to do explicit subgroup conjugacy tests. In some steps of the algorithm (say when calculating complements by cohomological methods) such tests can be preempted or reduced using other equivalences amongst the objects constructed.

In the following more detailed description, we shall focus on the task of finding all groups rather than the elimination of conjugates.

Methods similar to Sect. 5.1 can then be used to determine the incidence structure of the full subgroup lattice.

4 The Steps of the Algorithm

We now describe the different steps of the algorithm in more detail:

4.1 Factor Groups

A fundamental paradigm of the approach is to work in homomorphic images. This raises the question of how to represent factor groups of G in a suitable way. While this is difficult, in general, for the particular factor groups required here effective solutions exist:

- It has been shown [22, 35] that for permutation groups G, the factor $G/\mathrm{Rad}(G)$ can be (constructively) represented with permutation degree not exceeding that of G. (In GAP this is a call to `NaturalHomomorphismByNormalSubgroup (G,RadicalGroup(G))`. More generically, the special structure of $G/\mathrm{Rad}(G)$ as a subgroup of a direct product of wreath products allows for a representation of moderate degree, using imprimitive wreath products.
- By Schreier's conjecture (as proven in [20]), for a simple group T, the outer automorphism group $\mathrm{Aut}(T)/T$ is small. Thus $F/\mathrm{Soc}(F)$ (which embeds into a direct product of groups of the form $(\mathrm{Aut}(T_i)/T_i) \wr S_{m_i}$) is comparatively small and can be easily represented in an ad-hoc way.
- In many cases, it is not necessary to represent a factor group G/N faithfully, but it is sufficient to use representatives of elements and full preimages of subgroups. In particular, we can use this to perform linear algebra with coefficient vectors for the abelian factors R_i/R_{i+1} of the radical.

The question of the minimal permutation degree of factor groups of permutation groups have been studied also theoretically, and one can ask for other classes of normal subgroups for which such degree bounds hold:

Open Problem 2 *Extending the work of [18], describe (constructively) cases in which for permutation groups or matrix groups G and $N \lhd G$ one can represent the factor group G/N in degree not exceeding that of G.*

4.2 Subgroups of Simple Groups

Step 1 (from p. 4) asks us to determine the subgroups of a simple group T.

The basic method for this is the "cyclic extension" algorithm, dating back to [38]: A subgroup $S \leq T$ is either perfect, or there is a smaller subgroup $S' \leq U < S$ such that $S = \langle U, n \rangle$ with $n \in N_G(U)$. Thus:

(a) Initialize the perfect subgroups of T. This requires a precomputed list of isomorphism types of perfect groups such as [23] for groups of order at most 10^6. (By now, due to the rapid progress in computer engineering, the same methods would allow us to build such lists for larger orders.)
Then, in an approach close to isomorphism test algorithms, search for isomorphic copies of each of these groups as subgroups of T.
In GAP, the operation `RepresentativesPerfectSubgroups` can be used to obtain such a list.

(b) For every subgroup U listed so far, classify the U-orbits of elements of $N_G(U)$ outside U. If for an orbit representative n the group $\langle U, n\rangle$ is not yet known (i.e., not conjugate to a known group) then add it to the list. Iterate.

To allow for an efficient storage/comparison of subgroups, the algorithm maintains a list of cyclic subgroups of prime power order (called zuppos by their German acronym[2]). It then represents every subgroup as a bit list indicating which zuppos it contains.

Simple groups tend to have relatively few subgroups, enabling the calculation of subgroups even for large group orders. The assumed standard computer will calculate the subgroups of a simple group of order 10^5 in under a minute, order 10^6 about 5–10 min and (provided the potential perfect subgroups are available) order 10^7 about 90 min. (This is assuming that the group is given as a permutation group of minimal degree.)

The algorithm of course also will work for groups that are not simple, but in this case is often not competitive.

In GAP, this algorithm is implemented by the command `LatticeByCyclicExtension`.

In practice, we can (using this algorithm) create a database of subgroups of simple groups T up to a certain order limit once, and then store them. If the algorithm then is called for one of these simple groups, one then simply can fetch subgroups from the database.

GAP does exactly this, the databases used to obtain subgroup information is the library of tables of marks, provided by the `tomlib` package (which will be loaded automatically, if available). As of writing, this library contains full subgroup data for most of the simple groups in the ATLAS of order roughly up to 10^7. Some information about maximal subgroups of symmetric and alternating groups are also obtained through the library of primitive groups.

This approach requires an isomorphism between the concrete simple group T and its incarnation D in the database. Such an isomorphism can be facilitated in many cases through the use of so-called *standard generators* [44]: For a simple group T, this is a pair of elements $a, b \in T$ such that

- $T = \langle a, b\rangle$. (By [1] every finite simple group can be generated by two elements.)
- The pair (a, b) (that is its $\mathrm{Aut}(T)$-orbit) is characterized by simple relations, such as orders of a and b or short product expressions in a and b, or T-class memberships of a and b. This implies that if $T_1 \cong T_2 \cong T$ an isomorphism $T_1 \to T_2$ is obtained by finding instances of standard generators $a_1, b_1 \in T_1$ and $a_2, b_2 \in T_2$ and constructing the homomorphism that maps a_1 to a_2 and b_1 to b_2.
- In a given instance of T, such a pair (a, b) can be found quickly by only using basic group operations such as product and inverse (thus allowing for pseudo-random elements) and element order. A typical property achieving this is if the elements lie in small conjugacy classes that are powers of large conjugacy classes:

[2]"Zyklische Untergruppen von Primzahl-Potenz Ordnung".

A (pseudo-)random element will likely lie in a large class, by powering we get an element in the small class and only few conjugates to consider.

For example, $|a| = 2$, $|b| = 3$, $|ab| = 5$ could be used as such a generating set for A_5.

Such standard generators have been defined for all sporadic groups and many groups of Lie type of small order.

Open Problem 3 *Generalize "standard generators" to all quasisimple groups of Lie type.*

The concept of standard generators can be generalized to *constructive recognition*, that is the task to find an isomorphism from a simple group T to its stored database incarnation D, without relying on the need to find specific generators, but rather "rebuilding" natural combinatorial structures from within the group. For example, if the group T is a matrix group isomorphic to A_n, one might want to find a subspace of the natural module that has an orbit of length n under T, thus providing such an isomorphism through the action on the subspaces in the orbit. See the survey [16] for formal definitions and details.

4.3 Subdirect Products

Step 2 combines the subgroups of direct factors to those of a direct product. By induction, it is sufficient to consider the case of a direct product of two groups, $G \times H$. Let $S \leq G \times H$ and denote the projection from S to G by α and that from S to H by β. The image groups $A = S^\alpha$ and $B = S^\beta$ then are subgroups of G, respectively H.

Given such subgroups A and B, the construction of a *subdirect product* (which dates back at least to [41]) then allows to construct all groups S (see Fig. 2):

Denote by $D \lhd A$ the image of $\ker \beta$ under α and by $E \lhd B$ the image of $\ker \alpha$ under β. Then, by the isomorphism theorem

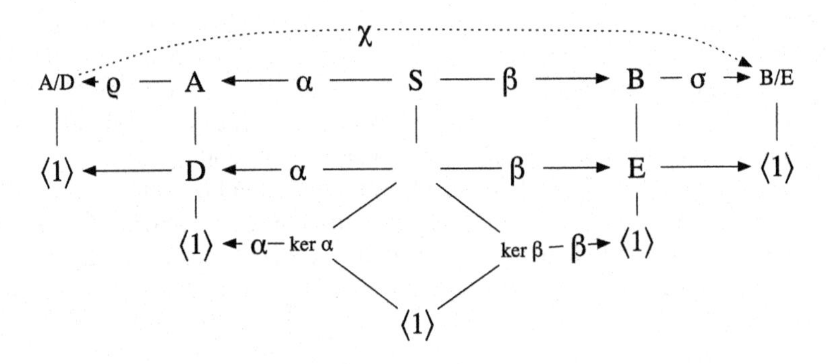

Fig. 2 Subdirect product construction

$$A/D \cong S/\langle \ker \alpha, \ker \beta \rangle \cong B/E.$$

If $\chi : A/D \to B/E$ is this isomorphism, and we denote the natural homomorphisms by $\varrho : A \to A/D$ and $\sigma : B \to B/E$, then

$$S = \{(a, b) \in G \times H \mid a \in A, b \in B, (a^{\varrho})^{\chi} = b^{\sigma}\}.$$

To construct all subdirect products S corresponding to the pair A, B, we thus classify pairs of normal subgroups $D \lhd A$, $E \lhd F$ together with isomorphisms $\chi : A/D \to B/E$.

Conjugacy of subgroups by $N_G(A) \times N_G(B)$ will induce equivalences on the normal subgroups and amongst the isomorphisms.

In the case we consider—subgroups of $\mathrm{Soc}(F)$—furthermore, there may be a conjugation action of F on the direct factors of its socle that causes further fusion of subgroups.

4.4 Normal Subgroups and Complements

In steps 4 and 6 of the calculation, we have a normal subgroup $N \lhd G$ and know the subgroups of G/N as well as the subgroups of N. (In step 6, the normal subgroup N is a vector space whose subgroups are easily enumerated.) From these we want to construct the subgroups of G.

We first analyze the situation: Let $S \leq G$ and set $A = \langle N, S \rangle$ and $B = S \cap N \lhd B$. (See Fig. 3, left.)

(A) Abelian Normal subgroup We consider first the case that N is abelian (which arises in step 6). Then $B \lhd N$ and thus $B \lhd \langle S, N \rangle = A$.

Thus, S/B is a complement to N/B in A/B. As N/B is elementary abelian, such complements can be obtained through cohomological methods, following [14].

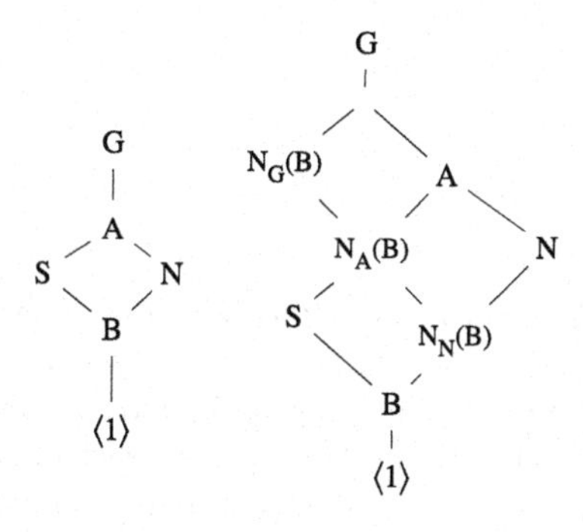

Fig. 3 Complement situations for subgroups

The input to such a computation is the linear action of A on N/B, together with a presentation for A/N.

To find all subgroups of G, we iterate through all A (as subgroups of G/N) and for each A determine candidates for B as submodules of N under the action of A [36].

As step 6 then iterates over a series, a crucial step towards efficiency is to extend a presentation for A/N to a presentation for S, if S/B is such a complement. This is easy, as B is elementary abelian.

(B) Nonabelian Normal subgroup If N is not abelian (as it will be in step 4), the situation is more complicated, as B is not necessarily normal in A, and there is no algorithm to easily determine complementing subgroups. In this case, following [27], we iterate through the possible subgroups $B \leq N$ and for each such B determine the groups S such that $S \cap N = B$:

As $N \leq \langle N, S \rangle = A$, we have that $N_N(B) \leq N_A(B) = \langle S, N_N(B) \rangle \leq N_G(B)$. Furthermore, $N_G(B)/N_N(B)$ is isomorphic to a subgroup of G/N. (See Fig. 3, right.) In this situation, S/B is a complement to $N_N(B)/B$ in $N_A(B)/B$.

Given a subgroup $B \leq N$, we thus determine the subgroups of $N_G(B)/N_N(B)$ (e.g., from the subgroups of G/N) and for each subgroup $N_A(B)/N_N(B)$ determine the candidates for S/B as complements. If $N_N(B)/B$ is solvable, this again can be done using cohomology calculations.

The group $N_N(B)/B$ does not need to be solvable—if the factor group, however, is solvable (which will be the case unless $\mathrm{Soc}(F)$ contains a single simple factor at least quintuply, in which case there will be storage problems already for the subgroups of $\mathrm{Soc}(F)$), [27] describes an approach for complements that reduces to p-groups, corresponding to a chief series of the factor.

GAP contains a function `ComplementClassesRepresentatives(G,N)` that determines representatives of the classes of complements to N in G, up to conjugacy by G, provided that N or G/N are solvable.

In the case that neither G and G/N are solvable, no algorithm for complements exists yet:

Open Problem 4 *Find a good algorithm for determining complements, if both normal subgroup and factor groups are not solvable. This also has relevance to maximal subgroup computations [10].*

4.5 Implementation

In GAP, the algorithm described in the previous sections is obtained through the operation `ConjugacyClassesSubgroups` (which used a number of variant methods, depending on the representation of the groups). It takes as argument a group and returns a list of conjugacy classes of subgroups. For each class `Representative` will return one subgroup; `AsList` applied to a class will return all subgroups in this class, thus

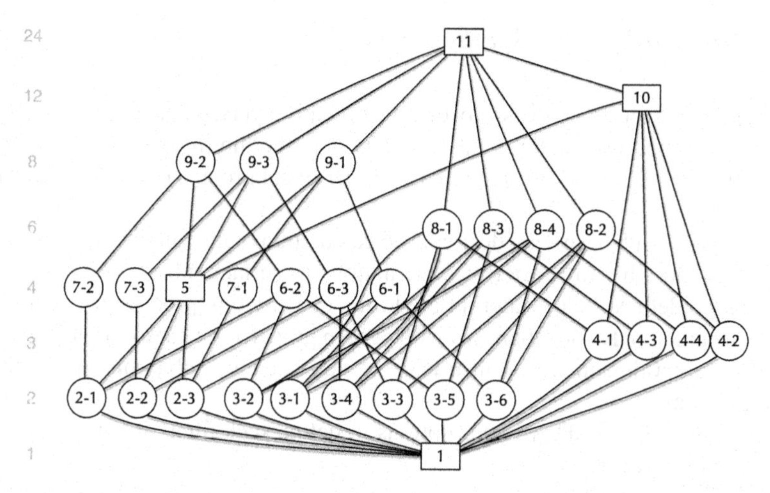

Fig. 4 Subgroup Lattice for S_4 (produced using `DotFileLatticeSubgroups` and then visualized using the graphics software OmniGraffle)

```
Concatenation(List(ConjugacyClassesSubgroups(G),AsList));
```

returns all subgroups of a group G. In general, such an enumeration of all subgroups are not recommended as it is very costly in terms of memory.

It is also possible to visualize the full lattice of subgroups of a group G. For this, the command

```
DotFileLatticeSubgroups(LatticeSubgroups(G),"filename.dot");
```

produces a text file, called `filename.dot` (or whatever file name is given) that describes the incidence structure of the subgroup lattice in the `graphviz` format (see www.graphviz.org for a description and for viewer programs for this format. There also are programs to convert this format into others, e.g., `dot2tex` converts to `TikZ` or `PSTricks` format.

Figure 4 illustrates the result in the example of the symmetric group S_4. Rectangles represent normal subgroups, circles ordinary subgroups and their conjugates. A number $a - b$ indicates group number b in class a (there is no b-part if the group is normal, as it will default to $b = 1$).

This group can be obtained in GAP then as `cl[`a`][`b`]` (that is `cl[8][3]` for $a = 8$ and $b = 3$) where `cl:=ConjugacyClassesSubgroups(G)`.

Caveat: The ordering (both a and b-parts) of subgroups can involve ad-hoc choices within the algorithm. When creating the group G a second time with the same generators, it is possible that a different numbering is chosen. It thus is not safe to use the $a - b$ indices for specifying a concrete subgroup outside a particular run of GAP.

4.6 Practicality and Modifications

With the construction process proceeding through layers, in each step proceeding through all subgroups found in the previous step, the limiting factor to calculation is (as timings in [27] indicate) the total number of subgroups, rather than the group order.

If only some subgroups are desired, and calculation of the full lattice is infeasible, it might be possible to restrict the calculations to certain subgroups, as long as a filter can be defined that is appropriate to the construction process and will iterate the construction only for subgroups with certain properties. (For example, the cyclic extension algorithm might be instructed to not calculate subgroups larger than a prescribed limit.)

At the moment, GAP provides options to define such filters in a few cases (see the manual for details):

- The operation `LatticeViaRadical` implements the general algorithm described here. If given two groups as argument it calculates subgroups of the second group up to conjugacy by the first group.
- `LatticeByCyclicExtension` allows for limiting the extension step to subgroups with a particular property.
- `SubgroupsSolvableGroup`, an implementation of the algorithm described for the case of solvable groups (in which case only step 6 is needed) allows to limit the determination of complements to specified cases, depending on properties of A, N, and B.
- In a different restriction, `SubgroupsSolvableGroup` also allows for determination of only those subgroups that are fixed (as subgroups) under a prescribed set of automorphisms, generalizing the concept of submodules [25].

5 Maximal, Low Index, and Intermediate Subgroups

A different class of algorithms is obtained by considering maximal subgroups.

If $M \le G$ is a maximal subgroup, the action of G on the cosets of M is primitive. The classification of primitive groups under the label O'Nan–Scott theorem [42] (see [34] for a full proof with corrections) thus can be used to describe possible maximal subgroups—one needs to search for quotient groups of G that have the correct structure to allow a primitive action, the point stabilizers for these actions will be maximal subgroups.

An approach to determine representatives of the conjugacy classes of maximal subgroups of a finite group, using this idea, is described in [11, 19]. The fundamental ingredients of these calculations again are the simple factors of $\mathrm{Soc}(F)$, and complements.

Taking again a series as described in Sect. 3, the algorithm then identifies factor groups of G that can have a faithful primitive representation. This is done via the socle of these subgroups, that is chief factors (or combinations of chief factors) of G:

- Maximal subgroups intersecting the radical lead to primitive actions of affine type and thus are obtained as complements. This is the only case of a solvable socle.
- Nonsolvable chief factors are obtained as part of $\mathrm{Soc}(F)$. Isomorphisms between the simple factors can be used to construct the different types of primitive actions, according to the diagonal and product action cases of the O'Nan–Scott theorem.
- The base case is maximal subgroups of simple groups, for which classifications exist in [30] and (far more explicitly) [7].

Open Problem 5 *Extend the concrete classification of maximal subgroups in [7] to larger degrees.*

As in the case of using stored tabulated information about subgroups, an explicit isomorphism needs to be found using constructive recognition or standard generators.

In GAP, representatives of the classes of maximal subgroups can be obtained using the function `MaximalSubgroupClassReps`. (Be aware that while `MaximalSubgroups` also exists, it will enumerate *all* maximal subgroups, often at significant cost.) Again tabulated information about maximal subgroups of simple groups is used.

5.1 Small Index and Intermediate Subgroups

The maximal subgroup functionality can be used to determine the maximal subgroups of a subgroup, thus obtaining maximal inclusion. (This also is used in general to provide the maximality relations required for the subgroup lattice structure.)

Iterating maximal subgroups (while avoiding recalculation of transferrable information) can be used to find subgroups that have bounded index [13]. One also could simply iterate the computation of maximal subgroups for all subgroups obtained so far to find subgroups that are k-step maximal in G. To reduce the cost, it will be natural to fuse conjugates under the action of the whole group.

In GAP, such latter functionality will be provided (starting with the 4.9 release) by a function `LowLayerSubgroups` that for a given group G and step limit k determines the subgroups of G, up to conjugacy, that is at most k-step maximal in G. It is possible to limit the calculation to obtain only subgroups of specified bounded index.

A further variant is to determine the *intermediate subgroups* $U < V < G$ for a given subgroup $U \leq G$ [29]: Instead of choosing an arbitrary representative M for each class of maximal subgroups, we determine in each step which conjugates of M contain the chosen subgroup U and then iterate.

This variant is implemented in GAP by the function `Intermediate` *Subgroups* (again this will see a significant performance improvement with the 4.9 release).

6 Summary

We have described the various methods that can be used in GAP to determine the subgroups of a given finite group. Different approaches provide different options to adapt the calculation. The methods also rely on a significant framework for basic operations that is essentially invisible to a user who does not look into the inner workings. While a calculation of subgroups is mostly limited by the size of the output set, there are still open research problems whose solution would improve this (and other) group-theoretic algorithms.

Acknowledgements The author's work has been supported in part by Simons Foundation Collaboration Grant 244502.

References

1. M. Aschbacher, R. Guralnick, Some applications of the first cohomology group. J. Algebra **90**(2), 446–460 (1984)
2. H. Bäärnhielm, D. Holt, C.R. Leedham-Green, E.A. O'Brien, A practical model for computation with matrix groups. J. Symbolic Comput. **68**(part 1), 27–60 (2015)
3. L. Babai, R. Beals, A polynomial-time theory of black box groups. I, in *Groups St Andrews 1997 in Bath* ed. by C.M. Campbell, E.F. Robertson, N. Ruskuc, G.C. Smith, vol. 260/261, London Mathematical Society Lecture Note Series (Cambridge University Press, Cambridge, 1999), pp. 30–64
4. L. Babai, I. Pak, Strong bias of group generators: an obstacle to the "product replacement algorithm". J. Algorithms **50**(2), 215–231 (2004). SODA 2000 special issue
5. L. Babai, A.J. Goodman, W.M. Kantor, E.M. Luks, P.P. Pálfy, Short presentations for finite groups. J. Algebra **194**, 97–112 (1997)
6. L. Babai, R. Beals, Á. Seress, Polynomial-time theory of matrix groups, in *Proceedings of the 41st Annual ACM Symposium on Theory of Computing, STOC 2009, Bethesda, MD, USA* (ACM Press, 2009), pp. 55–64
7. J.N. Bray, D.F. Holt, C.M. Roney-Dougal, *The maximal subgroups of the low-dimensional finite classical groups*, vol. 407 London Mathematical Society Lecture Note Series (Cambridge University Press, Cambridge, 2013). With a foreword by Martin Liebeck
8. J. Cannon, B. Souvignier, On the computation of conjugacy classes in permutation groups, in *Proceedings of the 1997 International Symposium on Symbolic and Algebraic Computation* ed. by W. Küchlin, The Association for Computing Machinery (ACM Press, 1997), pp. 392–399
9. J. Cannon, D. Holt, Automorphism group computation and isomorphism testing in finite groups. J. Symbolic Comput. **35**(3), 241–267 (2003)
10. J. Cannon, D. Holt, Computing maximal subgroups of finite groups. J. Symbolic Comput. **37**(5), 589–609 (2004)
11. J.J. Cannon, D.F. Holt, Computing conjugacy class representatives in permutation groups. J. Algebra **300**(1), 213–222 (2006)
12. J. Cannon, B. Cox, D. Holt, Computing the subgroup lattice of a permutation group. J. Symbolic Comput. **31**(1/2), 149–161 (2001)
13. J.J. Cannon, D.F. Holt, M. Slattery, A.K. Steel, Computing subgroups of bounded index in a finite group. J. Symbolic Comput. **40**(2), 1013–1022 (2005)
14. F. Celler, J. Neubüser, C.R.B. Wright, Some remarks on the computation of complements and normalizers in soluble groups. Acta Appl. Math. **21**, 57–76 (1990)

15. F. Celler, C.R. Leedham-Green, S.H. Murray, A.C. Niemeyer, E.A. O'Brien, Generating random elements of a finite group. Commun. Algebra **23**(13), 4931–4948 (1995)
16. H. Dietrich, C.R. Leedham-Green, E.A. O'Brien, Effective black-box constructive recognition of classical groups. J. Algebra **421**, 460–492 (2015)
17. J.D. Dixon, A. Majeed, Coset representatives for permutation groups. Portugal. Math. **45**(1), 61–68 (1988)
18. D. Easdown, C.E. Praeger, On minimal faithful permutation representations of finite groups. Bull. Austral. Math. Soc. **38**, 207–220 (1988)
19. B. Eick, A. Hulpke, Computing the maximal subgroups of a permutation group I, in *Proceedings of the International Conference at The Ohio State University, June 15–19, 1999*, vol. 8, ed. by W.M. Kantor, Á. Seress. Ohio State University Mathematical Research Institute Publications (Berlin, 2001), pp. 155–168. de Gruyter
20. W. Feit, Some consequences of the classification of finite simple groups, in *The Santa Cruz Conference on Finite Groups (Univ. California, Santa Cruz, Calif., 1979)*, vol. 37 *Proceedings of Symposia in Pure Mathematics* (American Mathematical Society, Providence, 1980), pp. 175–181
21. L. Finkelstein, W.M. Kantor, eds. *Groups and Computation II*, vol. 28, DIMACS: Series in Discrete Mathematics and Theoretical Computer Science (American Mathematical Society, Providence, 1997)
22. D.F. Holt, Representing quotients of permutation groups. Q. J. Math. Oxford Ser. (2) **48**(191), 347–350 (1997)
23. D.F. Holt, W. Plesken, *Perfect Groups* (Oxford University Press, Oxford, 1989)
24. D.F. Holt, B. Eick, E.A. O'Brien, *Handbook of Computational Group Theory*. Discrete Mathematics and its Applications (Chapman & Hall/CRC, Boca Raton, 2005)
25. A. Hulpke, Computing subgroups invariant under a set of automorphisms. J. Symbolic Comput. **27**(4), 415–427 (1999). (ID jsco.1998.0260)
26. A. Hulpke, Conjugacy classes in finite permutation groups via homomorphic images. Math. Comp. **69**(232), 1633–1651 (2000)
27. A. Hulpke, Calculation of the subgroups of a trivial-fitting group, in *ISSAC 2013—Proceedings of the 38th International Symposium on Symbolic and Algebraic Computation* (ACM, New York, 2013), pp. 205–210
28. A. Hulpke, Computing conjugacy classes of elements in matrix groups. J. Algebra **387**, 268–286 (2013)
29. A. Hulpke, Finding intermediate subgroups. Portugal. Math. **74**(3) (2017)
30. P. Kleidman, M. Liebeck, *The subgroup structure of the finite classical groups*, vol. 129 London Mathematical Society Lecture Note Series (Cambridge University Press, Cambridge, 1990)
31. R. Laue, J. Neubüser, U. Schoenwaelder, Algorithms for finite soluble groups and the SOGOS system, in *Computational group theory (Durham, 1982)* ed. by M.D. Atkinson (Academic press, 1984), pp. 105–135
32. J.S. Leon, Partitions, refinements, and permutation group computation, in *Finkelstein and Kantor [22]*, pp. 123–158
33. J.S. Leon, Permutation group algorithms based on partitions, I: theory and algorithms. J. Symbolic Comput. **12**, 533–583 (1991)
34. M.W. Liebeck, C.E. Praeger, J. Saxl, On the O'Nan-Scott theorem for finite primitive permutation groups. J. Austral. Math. Soc. Ser. A **44**, 389–396 (1988)
35. E.M. Luks, Á. Seress, Computing the fitting subgroup and solvable radical for small-base permutation groups in nearly linear time, in *Finkelstein and Kantor [21]*, pp. 169–181
36. K. Lux, J. Müller, M. Ringe, Peakword condensation and submodule lattices: an application of the meat-axe. J. Symbolic Comput. **17**, 529–544 (1994)
37. M. Mecky, J. Neubüser, Some remarks on the computation of conjugacy classes of soluble groups. Bull. Austral. Math. Soc. **40**(2), 281–292 (1989)
38. J. Neubüser, Untersuchungen des Untergruppenverbandes endlicher Gruppen auf einer programmgesteuerten elektronischen Dualmaschine. Numer. Math. **2**, 280–292 (1960)

39. M. Neunhöffer, Á. Seress, A data structure for a uniform approach to computations with finite groups, in *ISSAC 2006* (ACM, New York, 2006), pp. 254–261
40. E.A. O'Brien, Isomorphism testing for p-groups. J. Symbolic Comput. **17**, 133–147 (1994)
41. R. Remak, Über die Darstellung der endlichen Gruppen als Untergruppen direkter Produkte. J. Reine Angew. Math. **163**, 1–44 (1930)
42. L.L. Scott, Representations in characteristic p, in *The Santa Cruz conference on finite groups*, vol. 37, ed. by B. Cooperstein, G. Mason, *Proceedings of Symposia in Pure Mathematics* (American Mathematical Society, Providence, 1980), pp. 318–331. Corrigendum in [35]
43. Á. Seress, *Permutation Group Algorithms* (Cambridge University Press, Cambridge, 2003)
44. R.A. Wilson, Standard generators for sporadic simple groups. J. Algebra **184**(2), 505–515 (1996)

The Future of Majorana Theory

Alexander A. Ivanov

During most of my mathematical career, I was directly involved in the construction of the Monster group M, proof of its uniqueness, and understanding of its origins and structure. The study of the Monster group through its 2-local diagram geometry culminated in the simple connectedness proof for this geometry, reported in an invited 45-minute talk at the International Congress of Mathematicians in Kyoto in 1990 [14].

The simple connectedness result has played a key role in justification in my paper [15] and in S.P. Norton's [33] of the long-standing Y-conjecture, posed by B. Fischer and attacked by many prominent mathematicians including J.H. Conway, St. Linton, L.H. Soicher, and J. Tits. During the Durham Symposium on "Groups and Geometries" in July 1990, this conjecture became a theorem which John Conway called NICE (where "N" is for Norton, "I" is for Ivanov, "C" is for Conway, and "E" is for everyone else involved). The proof was the most spectacular example of the so-called "Geometric Presentations of Groups", the classical version of which is the Steinberg presentation for a group of Lie type.

The proof of the Y-conjecture made an important landmark in the theory of the Monster and gave a strong confidence in the power of the method of group amalgams in comparison with traditional tools. When I reported the proof of the Y-conjecture at an Oberwolfach meeting in early 1990s B. Fischer, present at this meeting, at first could not believe that such a proof was at all possible given the enormous size of the Monster group. In fact, the proof of the Y-conjecture was very conceptual in the sense that it applied the simple connectedness notion, familiar in algebraic topology to the study of groups defined via generators and relations.

Writing down the full proof of the simple connectedness result for the Monster geometry (personally inspired by John Thompson), including similar results for other geometries of sporadic groups and numerous consequences, took almost 20 years and resulted in a series of four monographs [16, 17, 19, 26]. The general procedure

---

A. A. Ivanov (✉)
Department of Mathematics, Imperial College, London, UK
e-mail: a.ivanov@imperial.ac.uk

N. S. N. Sastry and M. K. Yadav (eds.), *Group Theory and Computation*,
Indian Statistical Institute Series, https://doi.org/10.1007/978-981-13-2047-7_6

of the amalgam method involves two principal steps. On the first step, one constructs an amalgam which is going to generate the target group. In the case of the Monster, this step was accomplished in [18] and was praised by J. Tits in his personal letter as an ultimate accomplishment of the approach he proposed earlier. The second step involves analysis of cycles in the so-called Monster graph, which is a graph on the set of $2A$-involutions in the Monster, where two such involutions are adjacent whenever their product is again a $2A$-involution. A great deal of the structural information on the Monster graph was unearthed by S.P. Norton while proving the uniqueness of the Monster, which some consider as the final step in CFSG. The information in the uniqueness paper by S.P. Norton was sufficient to establish the first simple connectedness proof for the Monster in [14] but, in order to obtain a self-contained proof from basic principles it required a significant refinement. I always believed that in order to get hold of the cycles in the Monster graph one should consider the vertices as vectors in the Monster algebra. This expectation came true within the Majorana Theory. John Conway expressed his admiration for the Majorana Theory at the Conference in Princeton in his honor in 2015 after the plenary lecture given by me.

I always maintained close scientific cooperation with the Japanese mathematical community. These contacts started in 1983, when still being a postgraduate student I received a signed copy of the famous monograph by E. Bannai and T. Ito on algebraic combinatorics as an acknowledgment of a crucial contribution in [13] to the proof of the famous Bannai–Ito conjecture. In the supplement to the Russian translation of the monograph, published by me, E. Bannai and T. Ito referred to the main theorem of [13] as the "epoch-making result". By that time I was involved in the study of the Monster group, I had established close contacts with the school of M. Miyamoto working on Vertex Operator Algebras (VOAs). The discovery of Miyamoto involutions [29] was fascinating indeed and for many years I tried to deduce their existence directly from the properties of the Monster algebra. At a first glance, this looks like an impossible task, since the Miyamoto involution is deduced from the fusion rules of the Virasoro algebra of central charge one half. This is an infinite- dimensional algebra and the Monster algebra viewed as the algebra on the homogeneous elements at level 2 is just the tip of the iceberg. Meanwhile the amalgam approach to the Monster, developed in [15], enabled M. Miyamoto [31] to give a new construction of the Moonshine Module.

The situation gained a dramatic twist when in 2007 while on a sabbatical leave at the University of Tokyo, I learned from M. Miyamoto about a result of S. Sakuma [37], a former student of his. A detailed study of the rather short, but extremely condensed paper of Sakuma convinced me that Sakuma's result based on the fusion rule of a Virasoro algebra, could be fully translated in terms of the Monster algebra and that the properties of the algebra required for the proof have already been unearthed by J.H. Conway [4] and S.P. Norton [34]. This is the moment when the Majorana Theory was born. Back in London, the fusion rules were recognized by Alexander Gogolin (1965–2011), who was a great expert in the theory of elementary particles, as the fusion rules of the Majorana Fermion, this is how the name for the new theory was coined.

I maintained close mathematical cooperation with C.E. Praeger from the University of Western Australia for almost 20 years. This cooperation resulted in a number of joint publications including an exceptionally highly cited paper [23], which serves as a model for numerous further applications of the CFSG to algebraic graph theory, and an important paper [24] on locally projective graphs of girth 5. Recently, we have published a paper [10] on weakly locally projective graphs where the famous examples related to the sporadic groups M_{24} and He were characterized. It would be very important to extend this characterization to include the collinearity graph of the 2-local geometry of the Monster group, whose geometric girth 5.

1 The Tilde Geometry of the Monster

According to the Classification of Finite Simple Groups (CFSG), a finite non-abelian simple group is either an alternating group, a group of Lie type, or one of the 26 *sporadic* simple groups. The Lie-type groups enjoy a uniform theory of buildings and BN-pairs, while each of the sporadic simple groups has a story of its own. The fundamental question is "what is the purpose of the sporadic groups to exist" has both mathematical a philosophical flavor. For some groups, the answer can be found in J.Tits' brilliant survey [39] and in a later paper on the Monster group [40]. The Witt's design and the Golay code are the canonical tools to study the Mathieu groups. These structures pave the way to Conway groups through the Leech lattice and while Fischer's groups are best viewed as 3-transposition groups. The largest and most remarkable sporadic simple group, the Monster group was discovered independently by B. Fischer and R. Griess around 1973 and constructed by R. Griess in 1980 [11]. Michael Atiyah famously said that "the discovery of the Monster alone is the most exciting output of the classification of finite simple groups."

It is extremely desirable to locate the point, where sporadic groups diverged from the groups of Lie type. A Borel subgroup B related to the fixed prime p is the normalizer of a Sylow p-subgroup so it exists in any group G as long as p divides its order. Next one considers parabolic subgroups P_i where i runs through some index set. These are the subgroups containing B, and one considers the coset geometry $\mathcal{G}(G)$ of G associated with the subamalgam in G formed by the parabolic subgroups. If G is of Lie type and p is its natural characteristic then $\mathcal{G}(G)$ will be the corresponding building and its crucial feature is hidden in apartments stabilized by the Weyl subgroup, which, in turn, leads to the subgroup N in the BN-pair definition. The apartments keep the whole structure together, in particular, they force the coset geometry to be simply connected. This can be considered as a story with a happy end. For sporadic groups, the story is just the beginning.The parabolic subgroups can be defined as 2-local subgroups containing a given Sylow 2-subgroup, although the absence of apartments allows the parabolic subgroups to grow along with G, which might or might not be the universal completion of the amalgam.

The construction outlined in the previous paragraph applied to the Monster group leads to the tilde geometry $\mathcal{M}$ discovered by M. Ronan and St. Smith [36] and belonging to the following diagram:

$$\overset{1}{\underset{2}{\circ}} \overset{\sim}{=\!=} \overset{2}{\underset{2}{\circ}} \text{---} \overset{3}{\underset{2}{\circ}} \text{---} \overset{4}{\underset{2}{\circ}} \text{---} \overset{5}{\underset{2}{\circ}},$$

where the leftmost edge stands for the famous triple cover (known as the Foster graph) of the generalized quadrangle of order 2. The automorphism group of the generalized quadrangle of order 2 is the symplectic group $Sp_4(2)$, which is isomorphic to the symmetric group S_6 of degree 6, while the diagram is just the double edge. The tilde above this edge in the diagram of $\mathcal{G}(M)$ stands for the triple cover, whose automorphism group is the non-split extension $3 \cdot S_6$. The diagram without the tilde belongs to a unique geometry $\mathcal{G}(Sp_{10}(2))$. The point stabilizers in the latter classical geometry and in the Monster group are of the form

$$2^{10}.2^5.L_5(2) \text{ and } 2.2^5.2^5.2^{10}.2^{10}.2^5.L_5(2),$$

respectively.

The Monster group geometry $\mathcal{G}(M)$ was proved to be simply connected in [15] and it is an ongoing research to turn the simple connectedness result into a new completely self-contained construction for the Monster as was done for the Fourth Janko's group J_4 in [17]. The crucial step in the simple connectedness proof of $\mathcal{G}(M)$ is to prove that the fundamental group of the Monster graph is generated by the homology classes of the paths along the triangles. The Monster graph is a graph on the class of $2A$-involutions in the Monster, where two vertex-involutions are adjacent whenever their product is again a 2A-involution. In order to turn this into a construction of the Monster, one needs to show that $\Gamma(M)$ is uniquely recovered from its local structure. A major goal of my research is to accomplish this task, making essential use of the Majorana Theory described in the next section.

2 Majorana Algebras

Let V be a real vector space endowed with a non-associative commutative algebra product $\cdot$ and an inner product $(\ ,\ )$ which associate with each other, in the sense that $(u, v \cdot w) = (u \cdot v, w)$ for all $u, v, w \in V$. A vector $a \in V$ is said to be a *Majorana axis* if it is idempotent of length 1 and

(i) V is a direct sum of the eigenspaces of (the adjoint action of) a, every eigenvalue of a is in the set $\{1, 0, \frac{1}{4}, \frac{1}{32}\}$, and 1 is a simple eigenvalue;

(ii) the transformation $\tau(a)$ of V, which negates every $\frac{1}{32}$-eigenvector of a and fixes the remaining eigenvectors, preserves the algebra product;
(iii) the transformation $\sigma(a)$ of $C_V(\tau(a))$ which negates every $\frac{1}{4}$-eigenvector of a and fixes the 1- and 0-eigenvectors, preserves the restriction to $C_V(\tau(a))$ of the algebra product.

If V contains a set A of Majorana axes which generates V as an algebra, then $(V, A, \cdot, (\ ,\ ))$ is said to be a *Majorana algebra*. The algebra automorphism $\tau(a)$ as in (ii) is said to be a *Majorana involution*. If G is the isomorphism type of the subgroup in $GL(V)$ generated by the Majorana involutions $\tau(a)$ taken for all $a \in A$, then the natural homomorphism $\varphi : G \to GL(V)$ is said to be a *Majorana representation* of G.

The Majorana algebras and Majorana representations were introduced by me in [19] via axiomatization of some properties of the Monster. These properties were unearthed by J.H. Conway [4] and S.P. Norton [34], and proved in [19] as a step in constructing the Monster via group amalgams. Thus, it was shown in [19] that the famous 196,884-dimensional Conway–Griess–Norton algebra of the Monster is a Majorana algebra and that the action of the Monster on its algebra realizes a Majorana representation of the Monster. In this setting, the Majorana axes are the $2A$-axial vectors while the Majorana involutions are just the $2A$-involutions. In the context of Vertex Operator Algebras, the Majorana axes are conformal vectors of central charge one half and the Majorana involutions are the restrictions of the Miyamoto involutions to the homogeneous subalgebra at level 2. The theorem of Sakuma [37] gives the classification of the subalgebras generated by a pair of Majorana axes. The list of the possible subalgebras exactly matches such subalgebras in the Monster algebra, thus deducing the 6-transposition property of the $2A$-involutions in the Monster directly from the Majorana axioms.

The book [19] ignited a dramatic development which led to the formation of a totally new research area under the name *Majorana Theory*. Over a short period of time, through a number of important publications (cf. [1, 2, 6, 20, 21, 25, 27, 28, 38]) the theory gained its shape. An immediate outcome is an explicit construction of a number of important subalgebras in the Monster algebras, including two algebras of dimension 20 and 26 related to the $2A$-generated A_5-subgroups in the Monster [25]. The dimensions of these algebras were conjectured already in [19], although their identification could only be achieved within the Majorana Theory, since calculating in the whole of the 196,884-dimensional space is an impossible task.

The Majorana algebras constructed and classified so far, illuminate a remarkable tendency to embed into the Monster algebra. Some non-embeddable examples appear as Griess algebras of a new Vertex Operator Algebra (VOA), whose construction was motivated by the Majorana Theory [2]. A 70-dimensional Majorana algebra of the alternating group of degree 6 [21] is not embeddable into the Monster algebra and does not appear in any known VOA. This example, which probably could not be constructed outside of the Majorana Theory, demonstrates the diversity of the theory. The attempts to draw a clear borderline between the Monster embeddable and non-embeddable Majorana algebras inspired me to pose the following *Straight Flush Conjecture.*

Conjecture *Suppose that* $\mathcal{A}$ *is an indecomposable Majorana algebra in which for every* $i \in \{2, 3, 4, 5, 6\}$ *there exists a pair of Majorana involutions* τ_1 *and* τ_2*, such that the order of the product* $\tau_1\tau_2$ *is* i*. Then* $\mathcal{A}$ *embeds into the Monster algebra.*

A proof of the Straight Flush Conjecture will place the Monster algebra as the universal object in the class of Majorana algebras. The universality of the Monster will bring about a conceptual explanation of its numerous mysterious properties and eventually will provide an efficient tool for recognizing its subgroups and for performing transparent calculations with its elements.

Currently, we foresee a number of specific accomplishable goals toward the proof of the Straight Flush Conjecture, which have indisputable independent importance. These goals constitute the essential steps in attacking the conjecture and they are outlined in Sect. 4 after a brief review of the Monster algebra and its two-generated subalgebras in the next section.

3 The Monster Algebra and Norton–Sakuma Subalgebras

The Monster group M contains two conjugacy classes of involutions with representatives t and z, and respective centralizers

$$C \cong 2 \cdot BM \text{ and } D \cong 2_+^{1+24}.Co_1,$$

where BM is the Baby Monster sporadic simple group and Co_1 is the largest Conway sporadic simple group, whose double cover is the automorphism group of the Leech lattice. In a certain sense t resembles the behavior of a semi-simple element in an algebraic group, while z resembles that of a nilpotent element. The M-conjugates of t and z are called $2A$- and $2B$-involutions, respectively.

The minimal nontrivial complex representation of M has dimension 196 883 and it can be realized over the real numbers. It was noticed by Simon Norton, that (up to rescaling) the underlying vector space carries a unique M-invariant inner product and a unique M-invariant algebra product. The algebra product is non-associative, but it associates with the inner product. There is a special way of adjoining a trivial 1-dimensional submodule and rescaling the inner and algebra product to obtain a triple $(V, \cdot, (\ ,\ ))$, which is called the Conway–Griess–Norton algebra, or simply the Monster algebra of dimension 196 884.

It was shown in Chap. 8 of [19] that $(V, A, \cdot, (\ ,\ ))$ is a Majorana algebra, where A is the set of the $2A$-axial vectors. This enabled me to apply Sakuma's theorem

[37] to deduce that every Majorana algebra with two generating Majorana axes is embedded in the Monster algebras. Thus, the two-generated Majorana algebras are naturally indexed by the conjugacy classes in the Monster, which are products of pairs of $2A$-involutions:

$$2A,\ 2B,\ 3A,\ 3C,\ 4A,\ 4B,\ 5A,\ 6A.$$

The dimensions of these algebras are 3, 2, 4, 3, 5, 5, 6, 8, respectively. Inside the Monster, these algebras were calculated by S.P.Norton in [34] and in the abstract setting we call them Norton–Sakuma algebras. The important consequence of Sakuma's theorem is that a set of Majorana involutions is a 6-transposition set in the sense that the product of any two such involutions has order at most 6. Recall that the 6-transposition property of the $2A$-involutions in the Monster was the starting impulse of its discovery by B. Fischer.

4 Goals

Subalgebras with few generators. It is a natural question to ask about the structure of a subalgebra in a Majorana algebra generated by three Majorana axes, say a, b, and c. In complete generality the question is too complicated, since S.P. Norton has shown that the Monster algebra is 3-generated in many different ways. On the other hand, if a and b generate a $2A$-Sakuma–Norton algebra then by [32] the subgroup generated by the corresponding Majorana involutions τ_a and τ_b is a proper subgroup in the Monster (there are 27 such subgroups in the Monster up to conjugation). Madeleine Whybrow [41] has made a remarkable progress in the classification of the Majorana representations of such groups. She considered Majorana algebras generated by three Majorana axes a_0, a_1, and a_2 such that a_0 and a_1 generate a dihedral algebra of type 2A. We show that such an algebra must occur as a Majorana representation of one of 27 groups. These 27 groups coincide with the subgroups of the Monster which are generated by three 2A-involutions a, b, and c such that ab is also a 2A-involution, which were classified by S. P. Norton in [32]. Madeleine's work relies on that of S. Decelle [5] and consists of showing that certain groups do not admit Majorana representations. In all but one case, she has shown that the groups in question contain $2A$-pure elementary abelian subgroups of order eight which are well known not to admit a Majorana representation.

Inspired by the progress in the classification of the triangle subalgebras, we tried to attempt the case when a and b generate a $3A$-algebra. Although here the classification might be too complicated for the available techniques. The reason for this statement is that even a rough estimate gives a huge number of such configurations (even up to conjugation) in the Monster group. What we will try to do, is to classify the $(2A, 3A)$-configurations, which are subalgebras in a Majorana algebra generated by a Majorana axis and a $3A$-axis, the latter being the famous idempotent of squared length $8/5$ in the algebra of type $3A$. Inside the Monster algebra, these configurations are indexed by the pairs (t, u) where t is a $2A$-element, u is a $3A$-element (taken up to inversion). Simon Norton in [34] has listed all such pairs (up to conjugation) and it turned out that there are 22 of them. We aim to perform a similar classification in an arbitrary Majorana algebra. This requires the search of 6-transposition quotients of the modular group. This problem is certainly of great independent interest. To start, we will go through Norton's list of 22 pairs and study the subgroups in the Monster, generated by these pairs. It is another remarkable feature of the Monster that these groups are relatively small. The $3A$-axes satisfy the fusion rules of the 3-state Potts model described in [30] in the context of VOAs, where level 3 operators are explicitly involved. Lim Chien reformulated these rules in terms of Majorana axioms and used them to classify the $(2A, 3A)$-configurations in the Majorana algebra supporting the standard representation of A_{12} [3].

Majorana representations of groups. We start with a group G and address the existence question of a Majorana algebra $(V, (\ ,\ ), \cdot)$ and a set X of Majorana axes in V such that G is (isomorphic to) the group generated by the Majorana involutions $\tau(a)$ taken for all $a \in X$. Notice that G might act on the algebraic closure $AC(X)$ of X with a nontrivial kernel K contained in the center of G. If such an algebra exists, we say it is a *Majorana representation* of G. In these terms, Sakuma's theorem can be viewed as the classification of the Majorana representations of the dihedral groups. Since

$$\tau(a^{\tau(b)}) = \tau(b)^{-1}\tau(a)\tau(b),$$

it is natural to assume that $T := \{\tau(a) \mid a \in X\}$ is a union of conjugacy classes of involutions in G. When constructing a Majorana representation, at every stage we know some part of V, some values of the inner product and know how to multiply some vectors from the part we know. The procedure leads to a representation where the algebra product is shown to be closed. If the representation exhibits the uniqueness feature and there is an embedding $\varphi : G \to M$ of G into the Monster group such that $\varphi(T)$ consists of $2A$-involutions, we are able to conclude that V is precisely the subalgebra in the Monster algebra generated by the Majorana axes corresponding to $\varphi(T)$.

Thus, we have chosen a group G and a generating set T of its involutions such that $g^{-1}Tg = T$ for every $g \in G$. First, for every $t \in T$, we produce a vector a_t, which should become a Majorana axis with $\tau(a_t) = t$. Any two such vectors, say a_t and a_s generate a Norton–Sakuma algebra with the numerical part of the name equal to the order of ts. Thus, it is necessary for T to be a set of 6-transpositions. If the numerical part is 5 or 6, then the algebra is uniquely determined, otherwise there is a dichotomy. The assignment S, which maps each pair of elements of T to the isomorphism class of the Norton–Sakuma algebra they generate, is called the *shape* of the representation. This shape must respect the conjugation and the following inclusions among Norton–Sakuma algebras: $2A \subset 4B$, $2B \subset 4A$, $2A \subset 6A$, $3A \subset 6A$. When the shape is determined (or assigned), we know the set X and the restriction of $(\ ,\ )$ to X, in particular, we can determine the dimension of the linear span $LC(X)$ as the rank of the Gram matrix $\Gamma := ||(a_t, a_s)||_{t,s \in T}$. The matrix Γ must be positive definite since $(\ ,\ )$ is an inner product and the failure of this condition immediately demonstrates the nonexistence of the representation. If $LC(X) = AC(X)$ (so that the algebra product is closed on $LC(X)$) then we say that the algebra is 1-closed. The Monster algebra is 1-closed, although the Norton–Sakuma algebras $3A$, $4A$, $5A$, and $6A$ are not. In each of the four cases, it is sufficient to adjoin one specific vector to make the product closed, these vectors are known as $3A$, $4A$, $5A$, and $3A$ axes (recall that the $6A$-algebra contains the $3A$-algebra, so there are no such thing as "$6A$;-axes). By adjoining such axes for all Norton–Sakuma algebras corresponding to pairs from X and taking the linear span, we obtain the space $LC(X^2)$ which is called the 2-closure of X. If the algebra product is closed on $LC(X^2)$, the algebra is said to be 2-closed and this is the situation in which Á. Seress has made spectacular progress in [38].

Standard Representation of A_{12}. The Majorana representation of A_{12} with respect to the class T of bi-transpositions (which are products of pairs of commuting transpositions) has been at the center of our attention for a long time. There are a number of reasons for this including the fact that it is the largest ($2A$-generated) alternating subgroup in the Monster which contains most of the $2A + 1$ and $(2A, 3A)$-configurations. Many important properties of this representation were found in [1] including the fact that whenever the product of two bi-transpositions is a 3-cycle, the product of the corresponding Majorana axes is contained in $LC(A_{12})$. This fact was used in [8] to deduce that the algebra product in the Monster algebra is closed on the axes contained in the Harada–Norton group, so that the Majorana representation of the latter group is 1-closed. In [9], it was proved that 12 is the largest degree of an alternating group which possesses a Majorana representation with respect to the class of bi-transpositions. These results brought us close to the explicit description of the Majorana representation of A_{12} with respect to the class of bi-transposition. This will make a crucial break through in the whole Majorana Theory and will pave the way for a new independent construction of the Monster group. Therefore, this description is one of the main goals of our future research.

Weakly locally projective graphs. Returning to the geometrical foundation, we would like to obtain a stronger characterization of the collinearity graph of the 2-local parabolic geometry $\mathcal{G}(M)$ of the Monster in the class of weakly locally projective graphs of type $(5, 2, 2, 2)$ on the sense of [10]. It turns out that large Goldschmidt's amalgams (with Borel subgroup of order greater than 16) do not appear as the actions on planes and that the locally projective subgraphs can be defined in a functorial way. This is how the inevitability of the tilde geometry (along with the generalized quadrangle of order 2) will be established. A comparison of the classical examples against the Monster geometry will exhibit the Majorana algebras as non-associative analogs of the root system with Y-diagram being analogous to the Dynkin diagram.

The Mathieu Groups. It is planned to classify the Majorana representations of the five Mathieu groups M_{11}, M_{12}, M_{22}, M_{23}, and M_{24} and of their perfect central extensions. The former two groups are known to possess Majorana representations, while for the latter three the existence is questionable, since they are not $2A$-generated subgroups in the Monster and the latter two are not contained in the Monster at all. Although the Mathieu groups were around for a century and a half, and although there are more than five hundred papers dedicated to them, every time one needs some information about these group one has to take the challenge of performing own calculations. We took this chance to present a modern approach to the Mathieu groups based of the method of group amalgams [22]. The method which proved to be the best one when applied to the large sporadic simple groups including Baby Monster, the Fourth Janko Group, and the Monster, also works perfectly well for the Mathieu groups. The amalgam method as it crystallized by now involves a particular complex representation of the target group. Usually, this is the nontrivial representation of the smallest possible degree, but the main feature is the validity of the Thompson uniqueness criterium with respect to the amalgam under consideration. For the largest Mathieu group M_{24}, this is one of two complex conjugate 45-dimensional irreducible complex representations. The direct sum of the two conjugates obviously has dimension 90 and this is precisely the linear coefficient of the mock modular form associated with M_{24} by Tohru Eguchi [7] in a mysterious way. In [22], we include a self-contained construction of the 45-dimensional representations of M_{24}. Toward to the proof of the Straight Flush Conjecture, the knowledge of the Majorana representations of the Mathieu groups together with their extensions by outer automorphism and Schur multipliers is very important. Just to mention one of the features: according to Simon Norton, the smallest subalgebra of the Monster algebra, which accommodates all the eight types of Norton–Sakuma algebras is related to one of the Mathieu groups.

The axial algebras. The theory of axial algebras can be viewed as an offspring of the Majorana Theory which explores various generalizations of the fusions rules and leads to interesting theorems and examples (cf. [12], which is just the tip of the iceberg of preprints and papers).

References

1. A. Castillo–Ramirez, A.A. Ivanov, The axes of a Majorana representation of A_{12}, Springer Proceedings in Mathematics and Statistics, vol. 149, Ch. 9 (2014)
2. H.-Y. Chen, C.H. Lam, An explicit Majorana representation of the group $3^2 : 2$ of $3C$-pure type. Pacific J. Math. **271**, 25–51 (2014)
3. L. Chien, $3A$-axis in Majorana algebras, Ph.D. Thesis, Department of Mathematics, Imperial College (2017)
4. J.H. Conway, A simple construction for the Fischer-Griess monster group. Invent. Math. **79**, 513–540 (1985)
5. S. Decelle, Majorana Representations and the Coxerter Groups $G^{(m,n,p)}$, Ph.D. Thesis, Department of Mathematics, Imperial College London (2013)
6. S. Decelle, The $L_2(11)$-subalgebra of the Monster algebra. Ars Math. Contemp. **7**, 83–103 (2014)
7. T. Eguchi, H. Ooguri, Y. Tachikawa, Notes on the $K3$ surface and the Mathieu group M_{24}. Exp. Math. **20**, 9196 (2011)

8. C. Franchi, A.A. Ivanov, M. Mainardis, The $2A$-Majorana representations of the Harada-Norton group. Ars Math. Contemp. **11**(1), 175–187 (2016)
9. C. Franchi, A.A. Ivanov, M. Mainardis, Standard Majorana representations of the symmetric groups. J. Algebraic Combin. (2016)
10. M. Giudici, A.A. Ivanov, L. Morgan, C.E. Praeger, A characterisation of weakly locally projective amalgams related to A_{16} and the sporadic simple groups M_{24} and He. J. Algebra **460**, 340–365 (2016)
11. R.L. Griess, The friendly giant. Invent. Math. **69**, 1–102 (1982)
12. J.I. Hall, F. Rehren, S. Shpectorov, Primitive axial algebras of Jordan type. J. Algebra **437**, 79–115 (2015)
13. A.A. Ivanov, Bounding the diameter of a distance-regular graph. Soviet Math. Doklady **28**, 149–153 (1983)
14. A.A. Ivanov, Geometric presentation of groups with an application to the Monster, in *Proceedings of ICM-90, Kyoto, Japan, August 1990* (Springer, Berlin, 1991), pp. 385–395
15. A.A. Ivanov, A geometric characterization of the Monster, in *Groups, Combinatorics and Geometry, Durham 1990*, ed. by M. Liebeck, J. Saxl, London Mathematical Society Lecture Notes, vol. 165 (Cambridge University Press, Cambridge, 1992), pp. 46–62
16. A.A. Ivanov, *Geometry of Sporadic Groups. I. Petersen and Tilde Geometries*. Encyclopedia of Mathematics and its Applications, vol. 76 (Cambridge University Press, Cambridge, 1999), pp. xiv+408
17. A.A. Ivanov, *The Fourth Janko group*. Oxford Mathematical Monographs (The Clarendon Press, Oxford University Press, Oxford, 2004), xvi+233 pp
18. A.A. Ivanov, Constructing the Monster amalgam. J. Algebra **300**, 571–589 (2005)
19. A.A. Ivanov, *The Monster Group and Majorana Involutions*, vol. 176, Cambridge Tracts in Mathematics (Cambridge University Press, Cambridge, 2009)
20. A.A. Ivanov, On Majorana representations of A_6 and A_7. Comm. Math. Phys. **307**, 1–16 (2011)
21. A.A. Ivanov, Majorana representation of A_6 involving $3C$-algebras. Bull. Math. Sci. **1**, 365–378 (2011)
22. A.A. Ivanov, *The Mathieu Groups* (Cambridge University Press, Cambridge, 2018)
23. A.A. Ivanov, C.E. Praeger, On finite affine 2-arc transitive graphs. Europ. J. Comb. **14**, 421–444 (1993)
24. A.A. Ivanov, C.E. Praeger, On locally projective graphs of girth 5. J. Algebraic Combin. **7**(3), 259–283 (1998)
25. A.A. Ivanov, Á. Seress, Majorana representations of A_5. Math. Z. **272**, 269–295 (2012)
26. A.A. Ivanov, S.V. Shpectorov, *Geometry of Sporadic Groups. II. Representations and Amalgams.* Encyclopedia of Mathematics and its Applications, vol. 91 (Cambridge University Press, Cambridge, 2002), xviii+286 pp
27. A.A. Ivanov, S. Shpectorov, Majorana representations of $L_3(2)$. Adv. Geom. **12**, 717–738 (2012)
28. A.A. Ivanov, D.V. Pasechnik, Á. Seress, S. Shpectorov, Majorana representations of the symmetric group of degree 4. J. Algebra **324**, 2432–2463 (2010)
29. M. Miyamoto, Griess algebras and conformal vectors in vertex operator algebras. J. Algebra **179**, 523–548 (1996)
30. M. Miyamoto, 3-state Potts model and automorphisms of Vertex Operator Algebras of order 3. J. Algebra **239**, 56–76 (2001)
31. M. Miyamoto, A new construction of the moonshine vertex operator algebra over the real number field. Ann. Math.**59**, 535–596 (2004)
32. S.P. Norton, The uniqueness of the monster, in *Finite Simple Groups, Coming of Age*, ed. by J. McKay. Contemporary Mathematics, vol. 45 (AMS, Providence, 1985), pp. 271–285
33. S.P. Norton, Constructing the Monster, in *Groups, Combinatorics and Geometry, Durham 1990*, ed. by M. Liebeck, J. Saxl, London Mathematical Society Lecture Notes, vol. 165 (Cambridge University Press, Cambridge, 1992), pp. 63–76
34. S.P. Norton, The Monster algebra: some new formulae, in *Moonshine, the Monster and Related Topics*. Contemporary Mathematics, vol. 193 (AMS, Providence, 1996), pp. 297–306

35. S.P. Norton, Anatomy of the Monster I, in *The Atlas of Finite Groups: Ten Years On*, LMS Lecture Notes Series, vol. 249 (Cambridge University Press, Cambridge, 1998), pp. 198–214
36. M.A. Ronan, S. Smith, 2-Local geometries for some sporadic groups, in: *Proceedings of Symposia in Pure Mathematics*, No 37 (AMS, 1980), pp. 283–389
37. S. Sakuma, 6-Transposition property of τ-involutions of Vertex Operator Algebras, in *International Mathematics Research Notices* (2007), article rnm030, 19 p
38. Á. Seress, Construction of 2-closed M-representations, in *Proceedings of International Symposium on Symbolic and Algebraic Computation (ISSAC'12)* (2012), pp. 311–318
39. J. Tits, Groups finis simples sporadiques, Sèminaire N. Bourbaki, 1969–1970, exp. n° 375, pp. 187–211
40. J. Tits, On R. Griess' friendly giant. Invent. Math. **78**, 491–499 (1984)
41. M.A. Whybrow, Majorana algebras generated by a $2A$ algebra and a further axis. J. Group Theory [Submitted 2017]

Finite Groups with Abelian Automorphism Groups: A Survey

Rahul Dattatraya Kitture and Manoj K. Yadav

2010 Mathematics Subject Classification 20D45 · 20D15

1 Introduction

All the groups considered here are finite groups.

Any cyclic group has abelian automorphism group. By the structure theorem for finite abelian groups, it is easy to see that among abelian groups, only the cyclic ones have abelian automorphism groups. A natural question, posed by H. Hilton [19, Appendix, Question 7] in 1908, is:

Can a non-abelian group have abelian group of automorphisms?

An affirmative answer to this question was given by G. A. Miller [33] in 1913. He constructed a non-abelian group of order 2^6 whose automorphism group is elementary abelian of order 2^7. Observe that a non-abelian group with abelian automorphism group must be a nilpotent group of nilpotency class 2. Hence, it suffices to study the groups of prime power orders. Investigation on the structure of groups with abelian automorphism groups was initiated by C. Hopkins [20] in 1927. He, among other things, proved that such a group cannot have a nontrivial abelian direct factor, and if such a group is a p-group, then, so is its automorphism group, where p is a prime integer.

Unfortunately, the topic was not investigated for about half a century after the work of Hopkins. But days came when examples of such odd prime power order

R. D. Kitture · M. K. Yadav (✉)
School of Mathematics, Harish-Chandra Research Institute,
Chhatnag Road, Jhunsi, Allahabad 211019, India
e-mail: myadav@hri.res.in

R. D. Kitture
e-mail: rahul.kitture@gmail.com

R. D. Kitture · M. K. Yadav
Homi Bhabha National Institute, Training School Complex,
Anushakti Nagar, Mumbai 400085, India

N. S. N. Sastry and M. K. Yadav (eds.), *Group Theory and Computation*,
Indian Statistical Institute Series, https://doi.org/10.1007/978-981-13-2047-7_7

groups were constructed by D. Jonah and M. Konvisser [26] in 1975 and a thesis was written on the topic by B. E. Earnley [10] during the same year, in which, attributing to G. A. Miller, a group with abelian automorphism group was named as "Miller group". Following Earnley, we call a group to be *Miller*, if it is non-abelian and its automorphism group is abelian. Earnley pointed out that the former statement of Hopkins is not true as such for 2-groups. He constructed Miller 2-groups admitting a nontrivial abelian direct factor. However, the statement is correct for odd order groups. He also presented a generalization of examples of Jonah-Konvisser, which implies that the number of elements in a minimal generating set for a Miller group can be arbitrarily large. But he obtained a lower bound on the number of generators, which was later improved to an optimal bound by M. Morigi [34]. He also gave a lower bound on the order of Miller p-groups, which was again improved to an optimal bound by M. Morigi [35].

Motivated by the work done on the topic, various examples of Miller groups were constructed by several mathematicians via different approaches during the next 20 years, which were mostly special p-groups. A p-group G is said to be *special* if $Z(G) = G'$ is elementary abelian, where $Z(G)$ and G' denote the center and the commutator subgroup of G, respectively. These include the works by R. Faudree [12], H. Heineken and M. Liebeck [15], D. Jonah and M. Konvisser [26], B. E. Earnley [10], H. Heineken [16], A. Hughes [21], R. R. Struik [36], S. P. Glasby [13], M. J. Curran [7], and M. Morigi [35]. The existence of non-special Miller groups follows from [26, Remark 2].

The abelian p-groups of minimum order, which can occur as automorphism groups of some p-groups were studied by P. V. Hegarty [17] in 1995 and G. Ban and S. Yu [2] in 1998. More examples of Miller groups were constructed by A. Jamali [23] and Curran [8].

Neglecting Remark 2 in [26], A. Mahalanobis [30], while studying Miller groups in the context of MOR cryptosystems, conjectured that Miller p-groups are all special for odd p. Again neglecting [26, Remark 2], non-special Miller p-groups were constructed by V. K. Jain and the second author [24], V. K. Jain, P. K. Rai and the second author [25], A. Caranti [5] and the authors [28].

The motivation for many of the examples comes from some natural questions or observations on previously known examples. The construction of examples of Miller groups of varied nature has greatly contributed to complexify the structure of a Miller group. The structure of a general Miller group has not yet been well understood.

We record, here, the known information about structure of Miller p-groups and their automorphism groups. Let G be a Miller p-group, and assume that it has no abelian direct factor. Then, the following holds:

(1) $|G| \geq p^6$ if $p = 2$ and $|G| \geq p^7$ if $p > 2$.
(2) Minimal number of generators of G is 3 for p even, and 4 for p odd.
(3) The exponent of G is at least p^2.
(4) If $|G'| > 2$, then G' has at least two cyclic factors of the maximum order in its cyclic decomposition.

(5) If $\mathrm{Aut}(G)$ is elementary abelian, then $\Phi(G)$ is elementary abelian, $G' \le \Phi(G) \le Z(G)$ and at least one equality holds. There are Miller p-groups in which only one equality holds.
(6) For $p > 2$ (resp. $p = 2$), the abelian p-groups of order $< p^{12}$ (resp. $< 2^7$) are not automorphism groups of any p-group.

The aim of this article is to present an extensive survey on the developments in the theory of Miller groups since 1908, and pose some problems and further research directions. There are, at least, three survey articles on automorphisms of p-groups and related topics (see [18, 29, 32]); the present one does not overlap with any of them.

We conclude this section with setting some notations for a multiplicatively written group G. We denote by $\Phi(G)$, the Frattini subgroup of G. For an element x of G, $\langle x \rangle$ denotes the cyclic subgroup generated by x, and $o(x)$ denotes its order. For subgroups H, K of a group G, $H < K$ (or $K > H$) denotes that H is proper subgroup of K. The exponent of G is denoted by $\exp(G)$. By C_n, we denote the cyclic group of order n. For a p-group G and integer $i \ge 1$, $\Omega_i(G)$ denotes the subgroup of G generated by those x in G with $x^{p^i} = 1$, and $\mho^i(G)$ denotes the subgroup generated by x^{p^i} for all x in G. By $\mathrm{Inn}(G)$, $\mathrm{Autcent}(G)$ and $\mathrm{Aut}(G)$ we denote, respectively, the group of inner automorphisms, central automorphism, and all automorphisms of G. Let a group K acts on a group H by automorphisms, then $H \rtimes K$ denotes the semi-direct product of H by K. All other notations are standard.

2 Reductions

Starting from the fundamental observations of Hopkins, in this section, we summarize all the known results (to the best of our knowledge) describing the structure of Miller groups. Although we preserve the meaning, no efforts are made to preserve the original statements from the source.

As commented in the introduction, a non-abelian Miller group must be nilpotent (of class 2), and therefore, it is sufficient to study Miller p-groups. The following result is an easy exercise.

Proposition 1 (Hopkins, [20]) *Every automorphism of a Miller group centralizes* G'.

An automorphism of a group G is said to be *central* if it induces the identity automorphism on the central quotient $G/Z(G)$. Note that $\mathrm{Autcent}(G)$, the group of all central automorphisms of G, is the centralizer of $\mathrm{Inn}(G)$ in $\mathrm{Aut}(G)$.

A group G is said to be *purely non-abelian*, if it has no nontrivial abelian direct factor. Hopkins made the following important observation:

Theorem 2 (Hopkins, [20]) *A Miller p-group is purely non-abelian.*

Unfortunately, the statement, as such, is not true for $p = 2$ as shown by Earnley [10] (see Theorem 5 below for the correct statement).

In a finite abelian group G, there exists a minimal generating set $\{x_1, \ldots, x_n\}$ such that $\langle x_i \rangle \cap \langle x_j \rangle = 1$ for $i \neq j$. Hopkins observed that Miller groups possess a set of generators with a property similar to the preceding one. More precisely, he proved the following result.

Theorem 3 (Hopkins, [20]) *If G is a Miller p-group, then there exists a set $\{x_1, \ldots, x_n\}$ of generators of G such that*

(1) for $p > 2$, $\langle x_i \rangle \cap \langle x_j \rangle = 1$ for all $i \neq j$;
(2) for $p = 2$, $\langle x_i \rangle \cap \langle x_j \rangle$ is of order at most 2, for all $i \neq j$.

In fact, the theorem is true for *any p-group of class* 2, as shown below. Let G be a p-group of class 2. Choose a set $\{y_1, \ldots, y_k\}$ of generators of G with the property

$$\prod_i o(y_i) \text{ is minimum.} \qquad (*)$$

Let $p > 2$ and $o(y_i) = p^{n_i}$, $1 \leq i \leq k$. If $\langle y_1 \rangle \cap \langle y_2 \rangle \neq 1$ then $y_2^{p^{n_2-1}} \in \langle y_1 \rangle$. Assuming that $o(y_1) \geq o(y_2)$, we can write $y_2^{p^{n_2-1}} = (y_1^{p^{n_2-1}})^a$ for some integer a. Take $y_2' = y_2 y_1^{-a}$. Then, $\{y_1, y_2', y_3, \ldots, y_k\}$ is a generating set for G and $o(y_2') \leq p^{n_2-1} < o(y_2)$, which contradicts (*). The case $p = 2$ can be handled in a similar way.

Theorem 4 (Hopkins, [20]) *If G is a Miller p-group, then* $\mathrm{Aut}(G)$ *is a p-group.*

This can be proved easily in the following way. In the case when G is purely non-abelian, by a result of Adney-Yen [1, Theorem 1], $\mathrm{Autcent}(G)$ ($= \mathrm{Aut}(G)$) has order equal to $|\mathrm{Hom}(G, \mathrm{Z}(G))|$, which is clearly a power of p. Now, consider the case when G has an abelian direct factor, which occurs only when $p = 2$ and it must be cyclic of order at least 2^2 (see Theorem 5). Let $G = H \times C_{2^n}$, where H is purely non-abelian and $n \geq 2$. By the main theorem in [4],

$$|\mathrm{Aut}(G)| = |\mathrm{Aut}(H)| \cdot |\mathrm{Aut}(C_{2^n})| \cdot |\mathrm{Hom}(H, C_{2^n})| \cdot |\mathrm{Hom}(C_{2^n}, Z(H))|,$$

and each factor on the right side has order a power of 2.

The investigation of structure of Miller groups remained unattained for about half a century until it was revisited by Earnley [10] in 1974. He pointed out that a Miller 2-group *can* have an abelian direct factor. So, the correct form of Theorem 2 is

Theorem 5 (Earnley, [10]) *Let G be a finite p-group such that $G = A \times N$ with $A \neq 1$ an abelian group and N a purely non-abelian group. Then, G is a Miller group if and only if $p = 2$ and A, N satisfy the following conditions:*

(1) A is cyclic of order $2^n > 2$;
(2) N is a special Miller 2-group.

Theorem 6 (Earnley, [10]) *If G is a Miller p-group, then the following holds.*

(1) The exponent of G is greater than p;
(2) If $p > 2$, then $Z(G) \cap \Phi(G)$ is noncyclic.

For the first statement of the preceding theorem, we can assume that $p > 2$ and the proof now follows by noting that if $\exp(G) = p$, then the map $x \mapsto x^{-1}$ is a noncentral automorphism of G. The second statement follows from the following result of Adney-Yen.

Theorem 7 (Adney-Yen, [1]) *If G is a p-groups, $p > 2$, of class 2 such that G' has only one cyclic factor of maximum order in the direct product decomposition, then G possesses a noncentral automorphism.*

Since, in p-group of class 2, $G' \leq Z(G) \cap \Phi(G)$, the preceding theorem restricts the structure of the commutator subgroup of a Miller p-group.

Theorem 8 *If G is a Miller p-group, p odd, then G' possesses at least two cyclic factors of maximum order in the direct product decomposition.*

This raises a natural question for $p = 2$. The analog of the preceding theorem for $p = 2$ holds except when $|G'| = 2$. This can be obtained from the following generalization of Theorem 7.

Theorem 9 (Faudree, [11]) *Let G be a p-group of class 2 with the following conditions:*

(1) $G' = \langle u \rangle \times \mathrm{U}$, *where* $o(u) = p^{m_1} > p^{m'} = exp(\mathrm{U})$.
(2) $[g, h] = u$ *and* $h^{p^{m_1+m'}} = 1$,
(3) $m'' = m'$ *if p is odd, and* $m'' = max(1, m')$ *if* $p = 2$.

Let $H = \langle g, h \rangle$ *and* $L = \{x \in G : [g, x], [h, x] \in \mathrm{U}\}$. *Then* $G = HL$ *and the map*

$$g \mapsto gh^{p^{m''}}, \qquad h \mapsto h, \qquad x \mapsto x \ \ (x \in L)$$

defines an automorphism of G which centralizes $Z(G)$.

Notice that if $p = 2$ and the exponent of G' is at least 4, then the automorphism defined in the preceding theorem is noncentral, and therefore G is not Miller. But if $|G'| = 2$, then the theorem is no longer applicable to produce a noncentral automorphism of G. So, the following question remains open.

Question 1 Can a finite 2-group G with G' cyclic of order 2 be Miller?

Earnley obtained lower bounds for the order and the minimum number of generator of a Miller group, which were later sharpened to the optimal level by Morigi [34, 35]

Theorem 10 *Let G be a Miller p-group.*

(1) For any prime p, G is generated by at least 3 *elements. (Earnley, [10])*
(2) If p is odd, then G is generated by at least 4 *elements. (Morigi, [35])*

The example of a Miller 2-group constructed by Miller is minimally generated by 3 elements (see Sect. 3 (3.1)). For p odd, Jonah-Konvisser constructed Miller p-groups which are minimally generated by 4 elements (see Sect. 4 (4.3) for more details).

Theorem 11 *Let G be a Miller p-group.*

(1) For any prime p, $|G| \geq p^6$. (Earnley, [10])
(2) If p is odd, then $|G| \geq p^7$ and there exists a Miller group of order p^7. (Morigi, [34])

Again, the Miller 2-group constructed by Miller is of order 2^6 having automorphism group of order 2^7. Morigi constructed a special Miller p-group G of order p^7 with $\mathrm{Aut}(G)$ elementary abelian of order p^{12}. In fact, it is one among an infinite family of Miller groups constructed by Morigi (see Sect. 4 (4.6) for more details).

The question whether an abelian p-group of order smaller than p^{12} for an odd prime p can occur as the automorphism group of a p-group, was addressed by Hegarty [17] and Ban and Yu [2] independently. They proved

Theorem 12 *For p odd, there is no abelian p-group of order smaller than p^{12} which can occur as the automorphism group of a p-group.*

Finally, we state some results on the structure of $\mathrm{Aut}(G)$ for a Miller group G. Many known examples of Miller groups are special p-groups. Certainly, for such groups, the automorphism group is elementary abelian. The following problem appears as an *old problem* in [3, problem 722].

Problem 2 Study the p-groups with elementary abelian automorphism groups.

Let G be a p-group. If $\mathrm{Aut}(G)$ is elementary abelian, then so is $G/Z(G)$; hence, $G' \leq \Phi(G) \leq Z(G)$. In fact, it is interesting to see that at least one equality always holds.

Theorem 13 (Jain-Rai-Yadav, [25]) *Let G be a p-group, p odd, such that* $\mathrm{Aut}(G)$ *is elementary abelian. Then, $\Phi(G)$ is elementary abelian, and one of the following holds:*

(1) $Z(G) = \Phi(G)$.
(2) $G' = \Phi(G)$.

Jain-Rai-Yadav [25] constructed p-groups with elementary abelian automorphism group, in which exactly one of the above two conditions holds.

For 2-groups with elementary abelian automorphism groups, there are analogous necessary conditions, stated below. Since a Miller 2-group can have abelian (cyclic) direct factor, we consider two cases.

An abelian p-group of type $(p^n, p, \ldots, p)$ with $n > 1$ is called a *ce-group*. If $G = A \times B$ with $A \cong C_{p^n}$ $(n > 1)$ and $B \cong C_p \times \cdots \times C_p$, then call *A cyclic part* and *B elementary part* of G.

Theorem 14 (Jafari, [22]) *Let G be a purely non-abelian 2-group. Then,* $\text{Autcent}(G)$ *is elementary abelian if and only if one of the following holds.*

(1) G/G' *is of exponent* 2.
(2) $Z(G)$ *is of exponent* 2.
(3) $\gcd\big(\exp(G/G'), \exp(Z(G))\big) = 4$ *and* $G/G', Z(G)$ *are ce-groups such that elementary part of* $Z(G)$ *is contained in* G' *and there is an element* z *of order* 4 *in cyclic part of* $Z(G)$ *with* zG' *lying in cyclic part of* G/G' *satisfying* $o(zG') = \exp(G/G')/2$.

In particular, if G is a purely non-abelian 2-group with $\text{Aut}(G)$ elementary abelian, then one of the above three conditions holds. The example constructed by Miller shows that it satisfies only condition (3). Jain-Rai-Yadav [25] constructed purely non-abelian Miller 2-groups which satisfy only condition (1) or only condition (2). Finally, for 2-groups G with abelian direct factor, the following theorem gives necessary conditions for $\text{Aut}(G)$ to be elementary abelian.

Theorem 15 (Karimi-Farimani, [27]) *Let $G = A \times N$ with A cyclic 2-group and N purely non-abelian 2-group of class 2. Then,* $\text{Autcent}(G)$ *is elementary abelian if and only if*

(1) $|A|$ *is* 4 *or* 8.
(2) N *is special* 2*-group.*

With the notations of the preceding theorem, if G is Miller with $\text{Aut}(G)$ elementary abelian, then (1) and (2) hold.

Theorem 5 tells us that a Miller p-group, p odd, cannot admit a non-abelian direct factor. It is natural to ask whether a Miller p-group can occur as a direct product of non-abelian groups. This situation has been considered by Curran [8]. He determined necessary and sufficient conditions on a direct product $H \times K$ to have abelian automorphism group. Note that for any nontrivial group H, $\text{Aut}(H \times H)$ is non-abelian. Before we state the result of Curran, we set some notations.

Let H be a p-group of class 2.

(1) Let a, b, c, d denote the exponents of H/H', $H/Z(H)$, $Z(H)$, and H', respectively.
(2) If H' and $Z(H)$ have the same rank, then define d_s to be the largest integer ($\leq \exp(Z(H))$ such that $\Omega_{d_s}(H') = \Omega_{d_s}(Z(H))$.
(3) If $H/Z(H)$ and H/H' have the same rank, then define b_t to be the largest integer ($\leq \exp(H/H')$) such that $\Omega_{b_t}(H/Z(H)) \cong \Omega_{b_t}(H/H')$.
(4) Let $\Gamma^i(H)$ denote the subgroup of H with $\Gamma^i(H)/H' = \mho^i(H/H')$.

Replacing H by K in (1)–(4), the corresponding terms $a', b', c', d', d'_{s'}, b'_{t'}$ and $\Gamma^i(K)$ are similarly defined. For simplicity, we denote by $r(A)$, the rank of an abelian group A. With this setting, we have

Theorem 16 (Curran, [8]) *Let $G = H \times K$, where H, K are p-groups of class 2 with no common direct factor. Then* $\text{Aut}(G)$ *is abelian if and only if* $\text{Aut}(H)$ *and* $\text{Aut}(K)$ *are abelian and one of the following holds:*

(1) $Z(H) = H'$ *and* $Z(K) = K'$.
(2) $Z(H) > H'$ *and* $Z(K) = K'$, *where* $r(Z(H)) = r(H')$, $a' \leq d_s \leq a \leq c$ *and* $\Omega_a(Z(H)) \leq \Gamma^{c'}(H)$.
(3) $Z(H) = H'$ *and* $Z(K) > K'$, *where* $r(Z(K)) = r(K')$, $a \leq d'_{s'} \leq a' \leq c'$ *and* $\Omega_{a'}(Z(K)) \leq \Gamma^{c}(K)$.
(4) $Z(H) > H'$ *and* $Z(K) > K'$, *where* $r(Z(H)) = r(H')$, $r(Z(K)) = r(K')$ *and* $a = d_s = d'_{s'} = a'$.
(5) $Z(H) > H'$ *and* $Z(K) = K'$, *where* $r(H/Z(H)) = r(H/H')$, $a' \leq b_t \leq c \leq a$ *and* $\Omega_{a'}(Z(H)) \leq \Gamma^{c}(H)$.
(6) $Z(H) = H'$ *and* $Z(K) > K'$, *where* $r(K/Z(K)) = r(K/K')$, $a \leq b'_{t'} \leq c' \leq a'$ *and* $\Omega_a(Z(K)) \leq \Gamma^{c'}(K)$.
(7) $Z(H) > H'$ *and* $Z(K) > K'$, *where* $r(H/Z(H)) = r(H/H')$, $r(K/Z(K)) = r(K/K')$ *and* $c = b_t = b'_{t'} = c'$.

Remark 17 Observe that in the above theorem, if $\text{Aut}(H \times K)$ is abelian then either H' and $Z(H)$ have the same rank or H/H' and $H/Z(H)$ have the same rank; the same is true for the other component K.

An analogous problem for central product may be stated as follows.

Problem 3 Find necessary and/or sufficient condition such that central product of two Miller groups is again a Miller group.

3 Examples of Miller 2-Groups

In this section, we discuss examples of Miller 2-groups in chronological order.

(3.1) As mentioned above several times, the first example of a Miller 2-group was constructed by Miller himself, which comes as a semi-direct product of the cyclic group of order 8 by the dihedral group of order 8, and presented by

$$G_1 = C_8 \rtimes D_8 = \langle x, y, z \mid x^8, y^4, z^2, zyz^{-1} = y^{-1}, yxy^{-1} = x^5, zxz^{-1} = x \rangle.$$

Miller proved that each coset of $Z(G_1)$ is invariant under every automorphism of G_1 and that every automorphism has order dividing 2. Thus $\text{Aut}(G_1)$ is elementary abelian, and it can be shown that its order is 2^7.

By Theorem 11, the order of a Miller 2-group is at least 64. Having an example of order 64, a natural idea which peeps in ones mind is to explore groups of order 64 to find more Miller groups. This was done by Earnley [10], who proved that there are (exactly) two more Miller groups of the minimal order, which are presented as follows.

$$G_2 = (C_4 \rtimes C_4) \rtimes C_4 = \langle x, y, z \mid yxy^{-1} = x^{-1}, zxz^{-1} = xy^2, zyz^{-1} = y \rangle,$$

$$G_3 = (C_4 \times C_4 \times C_2) \rtimes C_2 = \langle x, y, z, t \mid x^4, y^4, z^2, t^2, xy = yx, xz = zx, yz = zy, \\ txt^{-1} = xy^2, tyt^{-1} = tz, t^{-2}zt = z \rangle.$$

Note that, in all the groups G_1, G_2 and G_3, the center and Frattini subgroups coincide and are elementary abelian of order 8, whereas the commutator subgroup is elementary abelian of order 4. Further, $\mathrm{Aut}(G_2)$ and $\mathrm{Aut}(G_3)$ are elementary abelian of order 2^9.

A generalization of G_1 appears, as an exercise, in the book [31, Exercise 46, Page 237] by Macdonald, written in 1970. For $n \geq 3$, the group is presented as follows.

$$G_{1,n} = \langle a, b, c \mid a^{2^n} = b^4 = c^2 = 1, b^{-1}ab = a^{1+2^{n-1}}, \quad c^{-1}bc = b^{-1}, \quad [c, a] = 1 \rangle.$$

The group $G_{1,n}$ is of order 2^{n+3} and its automorphism group is an abelian 2-group of type $(2^{n-2}, 2, 2, 2, 2, 2, 2)$. In 1982, Struik [36], independently, obtained the same example with a different presentation.

A generalization of G_2 and G_3 has also been obtained by Glasby [13], which is described as follows. For $n \geq 3$, let $G_{2,n}$ denotes the 2-group of class 2 with generators $x_1, x_2, \ldots, x_n$ with following additional relations:

$$x_i^4 = 1, \ (1 \leq i \leq n), \ [x_i, x_n] = x_{i+1}^2, \ (1 \leq i \leq n-1),$$

and set $[x_k, x_l] = 1$ in the remaining cases. Here, $\mathrm{Aut}(G_{2,n})$ is elementary abelian group of order 2^{2n}. If $n = 3$ then $G_{2,n}$ is isomorphic to G_2.

Again for $n \geq 3$, let $G_{3,n}$ denotes the 2-group of class 2 with generators y_0, y_1, ..., y_n with the following additional relations:

$$y_0^2 = y_i^4 = y_n^2 = 1, \ (1 \leq i \leq n-1), \ [y_i, y_n] = y_{i+1}^2, \ (1 \leq i \leq n-2)$$

and set $[y_k, y_l] = 1$ in the remaining cases. Here also $\mathrm{Aut}(G_{3,n})$ is elementary abelian group of order 2^{2n}. If $n = 3$, then $G_{3,n}$ is isomorphic to G_3.

(3.2) It should be noted that, in 1974, Jonah and Konvisser constructed special Miller groups of order p^8, which were generalized to an infinite family of Miller p-groups by Earnley, and those groups include the case $p = 2$ too (see Sect. 4 (4.3) and (4.4)).

(3.3) Heineken and Liebeck (see Sect. 4 (4.2) for details) proved that *given a finite group K, there exists a special p-group G, p odd, with* $\mathrm{Aut}(G)/\mathrm{Autcent}(G) \cong K$. In particular, if $K = 1$, then the corresponding group G is a Miller p-group.

In 1980, A. Hughes [21] proved that one can construct a special 2-group as well with the above property. The method of Hughes is a modification of the graph-theoretic method of Heineken-Liebeck. It should be noted that the method of Heineken-Liebeck uses *digraphs*, whereas that of Hughes considers *graphs*, and is described below.

Let K be any finite group and associate to K a connected graph $D(K)$ which satisfies the following conditions:

(1) Each vertex of the graph has degree at least 2.
(2) Every cycle in the graph contains at least 4 vertices.
(3) $\mathrm{Aut}(D(K)) \cong K$.

Associate a special 2-group G to $D(K)$ as follows. If the graph has n vertices $v_1, \ldots, v_n$, then consider the free group F_n on $x_1, \ldots, x_n$. Let R be the normal subgroup of F_n generated by x_i^2, $[x_i, [x_j, x_k]]$ (for all i, j, k) and $[x_r, x_s]$ whenever the vertices v_r and v_s are adjacent. Define G to be the group F_n/R; it is a special 2-group of order $2^{n+\binom{n}{2}-e}$, where n is the number of generators of G (so $|G/G'| = 2^n$) and e is the number of edges of the graph. It turns out that $\mathrm{Aut}(G)/\mathrm{Autcent}(G) \cong K$ (see [21] for proof).

(3.4) In 1987, Curran [7] studied automorphisms of semi-direct product, and suggested a method to construct many more examples of Miller 2-groups similar to G_1. We describe the method briefly and see that the above $G_{1,n}$ can be constructed by this method. Let $A = \langle a \rangle$ be a cyclic group of order 2^n, $n \geq 3$ and N a special 2-group acting on A in the following way: a maximal subgroup $J \leq N$ acts trivially, and any $x \in N \setminus J$ acts by

$$xax^{-1} = a^{1+2^{n-1}}.$$

Let $G = A \rtimes N$ be the semi-direct product of A and N with this action. Then, we get

Theorem 18 (Curran, [7]) *Let $G = A \rtimes N$ be as above along with the following conditions:*

(1) $A \times J$ is characteristic in $A \rtimes N$.
(2) Any automorphism of N leaving J invariant is central automorphism of N.

Then $A \rtimes N$ is a Miller group.

To elaborate, consider $A = \langle a \rangle$, the cyclic group of order 2^n, $n \geq 3$ and $N = \langle b, c \mid b^4, c^2, [b, c] = b^2 \rangle$, the dihedral group of order 8. Consider the action of N on A by

$$b^{-1}ab = a^{1+2^{n-1}}, c^{-1}ac = a.$$

Define $G = A \rtimes N$, the semi-direct product with this action. Note that, $Z(G) = \langle a^2, b^2 \rangle = \Phi(G)$, and $G/\Phi(G)$ is elementary abelian of order 8.

Take $J = \langle b^2, c\rangle$, the largest subgroup of N acting trivially on A. Then, $A \times J$ is characteristic in $A \rtimes N$, since it is the unique abelian subgroup of index 2 (if there were more than one abelian subgroups of index 2, then the center would have index 4). Also, in N, the subgroup $\langle b\rangle$ is characteristic. Hence, if $\varphi \in \mathrm{Aut}(N)$ leaves J invariant, then it leaves invariant the subgroup $J \cap \langle b\rangle = \langle b^2\rangle$ (the center of N), and one can see that φ is central automorphism of N. The group $A \rtimes N$ is, therefore, a Miller group by Theorem 18. Note that, this group is isomorphic to $G_{1,n}$ described above.

(3.5) After a considerable time gap, Jamali [23] in 2002 constructed the following infinite family of Miller 2-groups in which, the number of generators and the exponent of the group can be arbitrarily large. For integers $m \geq 2$ and $n \geq 3$, let $G_n(m)$ be the group generated by $a_1, \ldots, a_n, b$, subject to the following relations:

$$\begin{aligned}
&a_1^2 = a_2^{2^m} = a_i^4 = 1 \quad (3 \leq i \leq n)\\
&a_{n-1}^2 = b^2,\\
&[a_1, b] = [a_i, a_j] = 1 \quad (1 \leq i < j \leq n),\\
&[a_n, b] = a_1, [a_{i-1}, b] = a_i^2 \quad (3 \leq i \leq n).
\end{aligned}$$

The group $G_n(m)$ has order 2^{2n+m-2} and its automorphism group is abelian of type $(\underbrace{2, 2, \ldots, 2}_{n^2}, 2^{m-2})$. For $n = 3$ and $m = 2$, the group $G_3(2)$ is isomorphic to G_3 (see Sect. 3 (3.1)).

4 Examples of Miller p-Groups, p-Odd

This section, a lifeline for Miller groups in a sense, presents evolution of the topic. We will see the influence of examples of varied nature on Miller groups towards understanding the structure. Some examples of Miller groups occurred in other related contexts without any pointer to the topic. The deriving force behind many of the examples comes from natural questions on previously known Miller groups or certain natural optimistic expectations on the structure of such groups. The evolution process certainly helped, although minimally, in studying the structure of Miller groups, especially, in turning down the natural optimistic expectations.

During 1971–1979, there were three occasions, in which certain p-groups were constructed with a specific property, and these groups turned out to be Miller groups. The motive of the construction of these groups had no obvious connection with Miller groups.

(4.1) We now start the discussion of the (possibly) first example of Miller group of odd order. It was conjectured that a finite group in which every element commutes with its epimorphic image is abelian. It was disproved by R. Faudree [12] in 1971. He constructed a *non-abelian* group G in which every element commutes with its epimorphic image. The group G is a special p-group, and therefore one can easily deduce that $\mathrm{Aut}(G)$ is abelian. The group G is described as follows.

Let $G = \langle a_1, a_2, a_3, a_4 \rangle$ be the p-group of class 2 with the following additional relations:

$$[a_1, a_2] = a_1^p,\ [a_1, a_3] = a_3^p,\ [a_1, a_4] = a_4^p,$$
$$[a_2, a_3] = a_2^p,\ [a_2, a_4] = 1,\ [a_3, a_4] = a_3^p.$$

The group G is a special p-group of order p^8 and $Z(G) = G'$ is elementary abelian of order p^4. *If p is odd*, then $\mathrm{Aut}(G)$ is elementary abelian of order p^{16}.

(4.2) In 1974, Heineken-Liebeck [15], constructed a p-group G of class 2, p odd, for a given finite group K such that $\mathrm{Aut}(G)/\mathrm{Autcent}(G) \cong K$. Associate to the finite group K, a connected *digraph* (directed graph) X as follows. If K is cyclic of order > 2, then take X to be the cyclic digraph with $|K|$ vertices. In the remaining cases, we associate a digraph X with the following conditions:

(1) X is *strongly* connected,
(2) Any two non-simple vertices are *not* adjacent,
(3) Every vertex belongs to a (directed) cycle of length at least 5,
(4) Every non-simple vertex has at least two outgoing edges,
(5) $\mathrm{Aut}(X) \cong K$,

where by a *simple* vertex, we mean a vertex with exactly one incoming and exactly one outgoing edge. Associate to such X, a p-group G_X of class 2 as follows. If X has n vertices, we take n generators for G_X. If the vertex i has outgoing *edges* to $j_1, \ldots, j_r$ precisely, then we put the relation $x_i^p = [x_i, x_{j_1} \cdots x_{j_r}]$. Finally, we put $[[x_i, x_j], x_k] = 1$ for all i, j, k, making G_X of class 2. It is easy to show that G_X is a special p-group of order $p^{n+\binom{n}{2}}$.

The conditions (1)–(4) imply that $\mathrm{Aut}(G)/\mathrm{Autcent}(G)$ is isomorphic to the automorphism group of X, which is isomorphic to K (by (5)).

Before proceeding for the example of Miller p-group by this method, we make some comments. If $|K| \geq 5$, then there is a systematic procedure to construct graph X satisfying conditions (1)–(5); it is obtained by a specific subdivision of a Cayley digraph of K. If $|K| < 5$, then one constructs X satisfying (1)–(5) by some ad-hoc method.

For $K = 1$, the following digraph satisfies conditions (1)–(5).

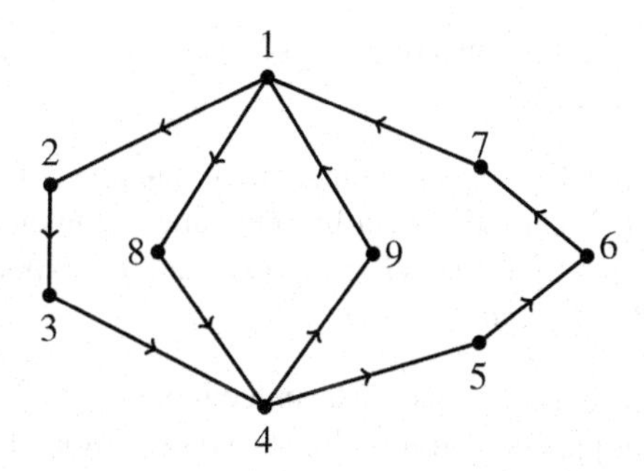

The group G_X associated to this digraph is a special Miller p-group of order $p^{9+\binom{9}{2}} = p^{45}$, p odd.

(4.3) The existence of Miller 2-groups of order 2^6 served as a base to the belief that p^{45} is too big to be a minimal order for a Miller p-group for odd p. This motivated Jonah and Konvisser [26] to construct Miller p-groups of smaller orders. In 1975, they constructed, for *each prime* p, $p+1$ non-isomorphic Miller p-groups of order p^8 as described below. Let $\lambda = (\lambda_1, \lambda_2)$ be a *non-zero* vector with entries in field $\mathbb{F}_p$ and $G_\lambda = \langle a_1, a_2, b_1, b_2 \rangle$ be the p-group of class 2 with the following additional relations:

$$a_1^p = [a_1, b_1],\ a_2^p = [a_1, b_1^{\lambda_1} b_2^{\lambda_2}],\ b_1^p = [a_2, b_1 b_2],$$
$$b_2^p = [a_2, b_2],\ [a_1, a_2] = [b_1, b_2] = 1.$$

The proof of the fact that G_λ is Miller group is very elegant. We briefly describe the idea here as it will be useful in later discussions.

Fix a nonzero vector λ and write $G = G_\lambda$. Note that, G is special p-group of order p^8 and G' is of rank 4 with generators $[a_i, b_j]$, $i, j = 1, 2$.

The subgroups $A = \langle a_1, a_2, G' \rangle$ and $B = \langle b_1, b_2, G' \rangle$ are the only abelian normal subgroups such that $[A : G'] = p^2 = [B : G']$; hence, they are permuted by every automorphism of G. There exists $x \in A$ such that $A^p \leq [x, G]$, but there is no $y \in B$ with $B^p \leq [y, G]$; hence, A, B are characteristic in G. The only $x \in A$ with $A^p \leq [x, G]$ are $a_1^i z$ with $p \nmid i$ and $z \in G'$; they generate characteristic subgroup $\langle a_1, G' \rangle$. Similarly, those $x \in A$ with $B^p \leq [x, G]$ generate a characteristic subgroup, namely $\langle a_2, G' \rangle$ and those $x \in B$ with $x^p \in [x, G]$ generate the characteristic subgroup $\langle b_2, G' \rangle$. Thus, if $\varphi \in \mathrm{Aut}(G)$, then

$$\varphi(a_1) \equiv a_1^i, \varphi(a_2) \equiv a_2^j, \varphi(b_1) \equiv b_1^{k_1} b_2^{k_2}, \varphi(b_2) \equiv b_2^l \pmod{G'},$$

where i, j, l are integers not divisible by p, and k_1, k_2 are integers, not simultaneously divisible by p. Using relations in G, it is now easy to deduce that $i \equiv j \equiv k_1 \equiv l \equiv 1 \pmod{p}$ and $k_2 \equiv 0 \pmod{p}$. Therefore, every automorphism of G is central. Since $G' = Z(G)$, it follows that $\mathrm{Aut}(G)$ is abelian.

The groups G_λ and G_μ, for nonzero vectors $\lambda, \mu \in \mathbb{F}_p \times \mathbb{F}_p$, are isomorphic only if these vectors are dependent.

Remark 19 Note that in the group G_λ, replacing, in the power relations, all the p-th power by p^k-th power, it can be shown by the same arguments that the resulting groups are still Miller. If $k > 1$, then $G'_\lambda = Z(G_\lambda)$ and has exponent $p^k > p$; hence G_λ is a *non-special Miller group*.

We must emphasize the negligence of the preceding remark [26, Remark 2] by the authors of many recent papers claiming the existence of non-special Miller groups was not known in the literature until recently.

(4.4) Examples of Jonah and Konvisser were generalized by Earnley [10], during the same year, by increasing number of generators, in the following way. Fix $n \geq 2$ and a nonzero n-tuple $(\lambda_1, \ldots, \lambda_n)$ with $\lambda_i \in \mathbb{F}_p$. Let $G_\lambda = \langle x_1, x_2, y_1, y_2, \ldots, y_n \rangle$ be the p-group of class 2 with the following additional relations:

$$x_1^p = [x_1, y_1], x_2^p = [x_1, y_1^{\lambda_1} \ldots y_n^{\lambda_n}],$$
$$y_i^p = [x_2, y_i y_{i+1}], i = 1, 2, \ldots, n-1,$$
$$y_n^p = [x_2, y_n], [x_1, x_2] = [y_i, y_j] = 1 \text{ for all } i, j.$$

It can be shown easily that $G'_\lambda = Z(G_\lambda)$ is elementary abelian and generated by the $2n$ elements $[x_i, y_j]$ for $i = 1, 2$ and $j = 1, 2, \ldots, n$; hence, order of G_λ is $p^{(n+2)+2n} = p^{2+3n}$.

For $\mathrm{Aut}(G_\lambda)$ to be abelian, the obvious necessary condition is that *every automorphism should be central*, but since G_λ is special, this is sufficient too. We briefly describe the beautiful linear algebra techniques evolved by Earnley to prove that every automorphism of G_λ is central as, with a little variation, these techniques have been used in different contexts by Morigi [35], Hegarty [17] and Earnley himself.

For simplicity, fix nonzero vector λ and write $G = G_\lambda$. Let $f : G/G' \to G'$ denote the map $xG' \mapsto x^p$, which is a homomorphism for odd p. An automorphism α of G/G' determines its action on G' by

$$\hat{\alpha} : G' \to G', \hat{\alpha}([x, y]) = [\alpha(xG'), \alpha(yG')].$$

The automorphism α of G/G' is induced by an automorphism φ of G if and only if $\hat{\alpha} \circ f = f \circ \alpha$, i.e., the following diagram commutes:

$$\begin{array}{ccc} G/G' & \xrightarrow{f} & G' \\ \downarrow \alpha & & \downarrow \hat{\alpha} \\ G/G' & \xrightarrow{f} & G'. \end{array}$$

Consequently, every automorphism of G is central if and only if identity is the only automorphism of G/G' which fits in the above commutative diagram.

Note that, G/G' and G' can be considered as vector spaces over $\mathbb{F}_p$; hence, the automorphisms $\hat{\alpha}$ and α in the above diagram can be considered as (invertible) linear maps. However, if $p = 2$, the map f *may not* be linear. But, in the group G under consideration, consider the *abelian* subgroups $A = \langle x_1, x_2, G' \rangle$ and $B = \langle y_1, \ldots, y_n, G' \rangle$; they generate G and the restriction of f to A/G' and B/G' are homomorphisms, and therefore linear.

Next, A and B are the only abelian subgroups containing G' such that modulo G' each of them have order at least p^2. For $n > 2$, both A and B are characteristic as their orders are different. But even for $n = 2$, they are characteristic, since, in this case, A contains an element t such that $A^p \leq [t, G]$, but B has no element with similar property.

Thus, consider an automorphism $\alpha = (\alpha_1, \alpha_2)$ of $G/G' = A/G' \oplus B/G'$, where $\alpha_1 \in \mathrm{Aut}(A/G')$ and $\alpha_2 \in \mathrm{Aut}(B/G')$. As a vector space, $G' = [A, B]$ is isomorphic to $A/G' \otimes B/G'$, the tensor product of A/G' and B/G'. Hence, the automorphism induced by α on G' is nothing but $\alpha_1 \otimes \alpha_2$. Then, α is induced by an automorphism of G if and only if the following diagrams commute:

$$\begin{array}{ccc} A/G' & \xrightarrow{f} & G' \\ {\scriptstyle\alpha_1}\downarrow & & \downarrow{\scriptstyle\alpha_1 \otimes \alpha_2} \\ A/G' & \xrightarrow{f} & G' \end{array} \qquad \begin{array}{ccc} B/G' & \xrightarrow{f} & G' \\ {\scriptstyle\alpha_2}\downarrow & & \downarrow{\scriptstyle\alpha_1 \otimes \alpha_2} \\ B/G' & \xrightarrow{f} & G' \end{array}$$

For the first diagram, consider the following ordered bases:

$$\{x_1G', x_2G'\} \text{ for } A/G' \text{ and } \{[x_1, y_1], [x_2, y_1], \ldots, [x_1, y_n], [x_2, y_n]\} \text{ for } G'.$$

Also for the second diagram, consider the following bases:

$$\{y_1G', \ldots, y_nG'\} \text{ for } B/G' \text{ and } \{[x_1, y_1], \ldots, [x_1, y_n], [x_2, y_1], \ldots, [x_2, y_n]\} \text{ for } G'.$$

The matrix of f with respect to these bases can be easily written from the *power-commutator relations* in G. Writing the matrices of $\alpha_1, \alpha_2, \alpha_1 \otimes \alpha_2$ with respect to these bases, one expresses the commutativity of above diagram in terms of two matrix equations, and a simple matrix computation shows that $\alpha_1 = 1$ and $\alpha_2 = 1$ are the only solutions. This implies that every automorphism of G is central. Note that, this also covers the case $p = 2$ as the restrictions of f to A/G' and B/G' are linear.

For more detailed module-theoretic formulation of the above arguments for arbitrary special p-groups, one may refer to [5].

(4.5) The third instance of examples of Miller groups, not in the context of the topic, is by Heineken [16] in 1979. He constructed a family of p-groups in which every normal subgroup is characteristic. The groups, actually, possess more interesting properties, which force the groups actually become Miller. The construction is as follows. Let $\mathbb{F}_q$ $(q = p^n)$ be the field of order p^n and $\mathrm{U}(3, \mathbb{F}_q)$ denote the group of 3×3 unitriangular matrices over $\mathbb{F}_q$. Identify the elements of $\mathrm{U}(3, \mathbb{F}_q)$ as triples over $\mathbb{F}_q$ by

$$(x, y, z) \leftrightsquigarrow \begin{bmatrix} 1 & x & z \\ & 1 & y \\ & & 1 \end{bmatrix}, x, y, z \in \mathbb{F}_q.$$

The sets $A = \{(x, 0, z) : x, z \in \mathbb{F}_q\}$ and $B = \{(0, y, z) : y, z \in \mathbb{F}_q\}$ constitute abelian subgroups of order $q^2 = p^{2n}$ and exponent p; they generate $\mathrm{U}(3, \mathbb{F}_q)$ and their intersection is the center of $\mathrm{U}(3, \mathbb{F}_q)$. Also, the commutator of an element of A with that of B is given by

$$[(x, 0, z_1), (0, y, z_2)] = (0, 0, xy).$$

Heineken constructed a group G by a little modification in the commutator and power relations of $\mathrm{U}(3, \mathbb{F}_q)$ as follows. The group G consists of triples (x, y, z) over $\mathbb{F}_q$, in which

$$A^* = \{(x, 0, z) \mid x, z \in \mathbb{F}_q\} \text{ and } B^* = \{(0, y, z) \mid y, z \in \mathbb{F}_q\}$$

constitute abelian subgroups of order $q^2 = p^{2n}$. The power relations are

$$(x, 0, z)^p = (0, 0, x^p - x),$$
$$(0, y, z)^p = (0, 0, y + y^p + \cdots + y^{p^{n-1}}).$$

For a fixed generator t of the multiplicative group of $\mathbb{F}_q$, the commutator relations between A^* and B^* are defined as

$$[(x, 0, z_1), (0, y, z_2)] = (0, 0, xy - tx^p y^{p^2}).$$

Then, $G = \{(x, y, z) \mid x, y, z \in \mathbb{F}_q\}$ is a group, which is the product of abelian subgroups A^* and B^*, and $Z(G) = \{(0, 0, z) \mid z \in \mathbb{F}_q\} = A^* \cap B^*$. Fix $x \neq 0$ and vary $y \in \mathbb{F}_q$, then the elements $xy - tx^p y^{p^2}$ exhaust whole $\mathbb{F}_q$. It follows that $G' = Z(G)$, and therefore G is a special p-group of order p^{3n}. With the above setup, we have

Theorem 20 *Let n be odd positive integer. If $n \geq 5$, $p > 2$ or $n = 3$, $p \geq 5$, then every automorphism of G is identity on $Z(G)$ as well as on $G/Z(G)$; hence,* $\mathrm{Aut}(G)$ *is abelian.*

The groups G in the preceding theorem have the following remarkable property: for every $x \in G - G'$, the conjugacy class of x in G is xG'. The groups satisfying this property are called *Camina* groups. The readers interested in Camina groups are referred to [9] and the references therein.

An automorphism α of a group is said to be *class-preserving* if it takes each element of the group to its conjugate. It is an easy exercise to show that each automorphism of the group G is class-preserving. This is not only true for the groups G considered above, but also for any Camina p-group which is Miller as well.

Theorem 21 *Let G be a p-group which is Miller as well as a Camina group. Then automorphisms of G are all class-preserving.*

As a simple consequence, we have

Corollary 22 *Let G be a p-group which is Miller as well as a Camina group. Then normal subgroups of G are all characteristic.*

We remark that the examples of Jonah-Konvisser are not Camina groups and extra special p-groups are Camina groups but not Miller. As the structure of Camina p-groups of class 2 in general is not well understood, it will be interesting to study p-groups which are both Camina as well as Miller.

Problem 4 Determine the structure of p-groups which are Camina as well as Miller.

This structural information will also, on the one hand, help in understanding p-groups of class 2 whose automorphisms are all class-preserving and, on the other hand, shed some light on the study of p-groups of class 2 in which all normal subgroups are characteristic.

(4.6) Note that, the Miller groups constructed by Heineken in the preceding discussion are of order at least p^9. But, as we already mentioned, the lower bound on the order of Miller groups is p^7 for odd p. This was Morigi [35], who, in 1994, constructed examples of Miller groups of minimal order as a part of a general construction of an infinite family of such groups. The construction is briefly described as follows. For any natural number n, let G^n denote the p-group of class 2 generated by $a_1, a_2, b_1, \ldots, b_{2n}$ with the following additional commutator and power relations:

$$\begin{aligned}
&[a_1, b_{2i+1}] = [a_2, b_{2i+2}] = 1, i = 0, 1, \ldots, n-1,\\
&[b_1, b_2] = [b_3, b_4] = \cdots = [b_{2n-1}, b_{2n}] = 1,\\
&[b_i, b_j] = 1 \text{ if } i \equiv j \pmod 2.
\end{aligned}$$

Further, $a_1^p = a_2^p = 1$ and b_1^p are the product of the following $n^2 + n + 1$ commutators:

$$[a_1, a_2], [a_1, b_{2i+2}], [a_2, b_{2i+1}], [b_{2i+1}, b_{2j+2}], i, j = 0, 1, \ldots, n-1, i \neq j. \quad (**)$$

Finally, powers of b_i's are related by

$$b_2^p = b_1^p[a_1, b_2]^{-1}$$
$$b_{2i+1}^p = b_{2i}^p[a_2, b_{2i+1}]^{-1}, \quad b_{2i+2}^p = b_{2i+1}^p[a_1, b_{2i+2}]^{-1}, i = 1, \ldots, n-1.$$

Then, G^n is a special p-group, with $|G^n/(G^n)'| = p^{2n+2}$ and $(G^n)'$ is elementary abelian p-group generated by n^2+n+1 commutators in (**); so $|G^n| = p^{n^2+3n+3}$. Further, $\mathrm{Aut}(G^n)$ is (elementary) abelian, and

$$|\mathrm{Aut}(G^n)| = |\mathrm{Autcent}(G^n)| = p^{(2n+2)(n^2+n+1)}.$$

For $n = 1$, G^1 is a special p-group of order p^7. This is an example of a Miller group of the smallest order for p odd. The following problem is interesting:

Problem 5 Describe all Miller p-groups of order p^7 for $p > 2$.

(4.7) Up to this point, we have only considered the examples of special Miller p-groups (modulo Remark 19). Now, we will consider non-special groups. Before we proceed further with more examples, we record a very useful result of Adney and Yen [1].

Let G be a purely non-abelian p-group of class 2. Let $G/G' = \prod_{i=1}^r \langle x_i G' \rangle$, with $o(x_i G') \geq o(x_{i+1} G')$. Let p^a, p^b, p^c denote the exponents of $Z(G)$, G' and G/G', respectively. Finally, let R be the subgroup of $Z(G)$ generated by all the homomorphic images of G in $Z(G)$, and K denote the intersection of the kernels of all the homomorphisms $G \to G'$.

Theorem 23 (Adney-Yen, [1]) *With the above notations,* $\mathrm{Autcent}(G)$ *is abelian if and only if the following holds:*

(1) $R = K$.
(2) either $\min(a, c) = b$ *or* $\min(a, c) > b$ *and* $R/G' = \langle x^{p^b} G' \rangle$.

Motivated by the conjecture of Mahalanobis as stated in the introduction, Jain and the second author [24], in 2012, constructed the following infinite family of non-special Miller p-groups. For $n \geq 2$, and p odd, let $G_n = \langle x_1, x_2, x_3, x_4 \rangle$ be the p-group of class 2 with the following additional relations:

$$x_1^{p^n} = x_2^{p^2} = x_3^{p^2} = x_4^p = 1,$$
$$[x_1, x_2] = x_2^p, \ [x_1, x_3] = [x_1, x_4] = x_3^p,$$
$$[x_2, x_3] = x_1^{p^{n-1}}, \ [x_2, x_4] = x_2^p, \ [x_3, x_4] = 1.$$

It is then easy to see that

(1) $Z(G_n) = \Phi(G_n) = \langle x_1^p, x_2^p, x_3^p \rangle$.
(2) $G_n' = \langle x_1^{p^{n-1}}, x_2^p, x_3^p \rangle$ is elementary abelian of order p^3.
(3) G_n is special only when $n = 2$ (follows by (1) and (2)).

The proof that $\text{Aut}(G_n) = \text{Autcent}(G_n)$ is constructive and reply on detailed careful calculations. An application of Theorem 23 now shows that G_n is Miller.

In the preceding examples of non-special Miller p-groups G, one can easily notice that

$$G' < Z(G) = \Phi(G).$$

One might desire that for a Miller p-group G, one of the following always holds true: (i) $G' = Z(G)$; (ii) $Z(G) = \Phi(G)$.

In 2013, Jain, Rai and the second author [25] constructed the following infinite family of Miller p-groups G such that $G' < Z(G) < \Phi(G)$, which we again denote by G_n. For $n \geq 4$, let $G_n = \langle x_1, x_2, x_3, x_4 \rangle$ be a p-group of class 2 with the following additional relations:

$$\begin{aligned}
&x_1^{p^n} = x_2^{p^4} = x_3^{p^4} = x_4^{p^2} = 1\\
&[x_1, x_2] = [x_1, x_3] = x_2^{p^2},\ [x_1, x_4] = x_3^{p^2},\\
&[x_2, x_3] = x_1^{p^{n-2}},\ [x_2, x_4] = x_3^{p^2},\ [x_3, x_4] = x_2^{p^2}.
\end{aligned}$$

Then, G_n is a p-group of order p^{n+10} with

$$Z(G_n) = \langle x_1^{p^2}, x_2^{p^2}, x_3^{p^2} \rangle,\ \Phi(G_n) = \langle x_1^p, x_2^p, x_3^p, x_4^p \rangle,\ G_n' = \langle x_1^{p^{n-2}}, x_2^{p^2}, x_3^{p^2} \rangle.$$

It follows that $G_n' \leq Z(G_n) < \Phi(G_n)$ and $G_n' = Z(G_n)$ only when $n = 4$. As in the previous examples, the proof of the fact that G_n is Miller is constructive.

As mentioned in Theorem 13, if $\text{Aut}(G)$ is elementary abelian, then $\Phi(G)$ is elementary abelian and one of the following holds: (1) $Z(G) = \Phi(G)$; (2) $G' = \Phi(G)$. Jain, Rai and the second author constructed the following Miller p-groups in which exactly one of these conditions hold.

For any prime p, let $G_4 = \langle x_1, x_2, x_3, x_4 \rangle$ denote the p-group of class 2 with the following additional relations:

$$\begin{aligned}
&x_1^{p^2} = x_2^{p^2} = x_3^{p^2} = x_4^{p^2} = 1,\\
&[x_1, x_2] = 1,\ [x_1, x_3] = x_4^p,\ [x_1, x_4] = x_4^p,\\
&[x_2, x_3] = x_1^p,\ [x_2, x_4] = x_2^p,\ [x_3, x_4] = x_4^p.
\end{aligned}$$

Then, G_4 is a p-group of order p^8 and the following holds.

(1) $G_4' = \langle x_1^p, x_2^p, x_4^p \rangle$ is elementary abelian of order p^3.
(2) $\Phi(G_4) = Z(G_4) = \langle x_1^p, x_2^p, x_3^p, x_4^p \rangle$ is elementary abelian of order p^4.
(3) $\text{Aut}(G_4)$ is elementary abelian of order p^{16}.

Again for any prime p, consider the p-group $G_5 = \langle x_1, x_2, x_3, x_4, x_5 \rangle$ of class 2 with the following additional relations:

$$x_1^{p^2} = x_2^{p^2} = x_3^{p^2} = x_4^{p^2} = x_5^p = 1,$$
$$[x_1, x_2] = x_1^p,\ [x_1, x_3] = x_3^p,\ [x_1, x_4] = 1,\ [x_1, x_5] = x_1^p,\ [x_2, x_3] = x_2^p,$$
$$[x_2, x_4] = 1,\ [x_2, x_5] = x_4^p,\ [x_3, x_4] = 1,\ [x_3, x_5] = x_4^p,\ [x_4, x_5] = 1.$$

Then, G_5 is a p-group of order p^9 and the following holds.

(1) $G_5' = \Phi(G_5) = \langle x_1^p, x_2^p, x_3^p, x_4^p \rangle$ is elementary abelian of order p^4.
(2) $Z(G_5) = \langle x_4, G_5' \rangle$.
(3) $\mathrm{Aut}(G_5)$ is elementary abelian of order p^{20}.

It is natural to ask

Question 6 Does there exist a Miller p-group G in which $Z(G) \nsubseteq \Phi(G)$ and $\Phi(G) \nsubseteq Z(G)$?

(4.8) The proofs of the results in Sect. (4.7) involve heavy computations. To remedy the problem, in 2015, Caranti [5, 6] suggested a simple module- theoretic approach to construct non-special Miller p-groups from special ones. The arguments given in [5] are not sufficient to prove the results as stated. The authors of the present survey proved that the results are valid under an additional hypothesis. The construction is briefly described as follows.

For an odd prime p, let H be a special Miller p-group satisfying the following hypotheses:

(i) $\mho_1(H)$ is a proper subgroup of H'.
(ii) The map $H/H' \to H'$ defined by $hH' \mapsto h^p$ is injective.

Let $K = \langle z \rangle$ be the cyclic group of order p^2, and M be a subgroup of order p in H' but not in $\mho_1(H)$. Let G be a central product of H and K amalgamated at M. Note that $G' = \Phi(G) < Z(G)$; hence G is non-special. With this setting, we have

Theorem 24 *If H/M is a Miller group, then so is G.*

Before we proceed, we make a comment on the preceding theorem. Caranti claimed that G is Miller without the condition "H/M is Miller". Unfortunately, this is not always true, as shown in the following example.

Let $H = \langle a, b, c, d \rangle$ be the p-group of class 2 with the following additional relations:

$$a^p = [a, c], \quad b^p = [a, bcd], \quad c^p = [b, cd], \quad d^p = [b, d].$$

Then, H is a special Miller p-group of order p^{10} and satisfies conditions (i)–(ii). It can be proved that if $M = \langle [a, b] \rangle$, then G is Miller, and if $M = \langle [a, d] \rangle$, then G is not a Miller group. That H and $H/\langle [a, b] \rangle$ are Miller can be proved following the arguments similar to those in [26] (for details see [28]).

As noted above, the Miller groups G are such that $G' = \Phi(G) < Z(G)$. Now with a little variation in the preceding construction, we obtain Miller p-groups G with $G' < \Phi(G) = Z(G)$. Assume that the special Miller group H also satisfies the following hypothesis in addition to (i)–(ii) above:

(iii) If H is minimally generated by $x_1, \ldots, x_n$, then H' is minimally generated by $[x_i, x_j]$ for $1 \leq i < j \leq n$.

Let $L = \langle z \rangle$ be a cyclic group of order p^n, $n \geq 3$ and let z act on H via a non-inner central automorphism σ of H (which always exists in a Miller p-group). Let N be a subgroup of order p in H' but not in $\mho_1(H)$. Now define G to be the *partial semi-direct product* of H by L amalgamated at N (cf. [14] or [28]). With this setting, we finally have

Theorem 25 *If H/N is a Miller group, then so is G.*

Again, we remark that the condition "H/N is Miller", in the preceding theorem, cannot be dropped, as shown by the following example. Let $H = \langle a, b, c, d \rangle$ be the p-group of class 2 described above. Then, the map σ defined by

$$a \mapsto ad^p, b \mapsto b, c \mapsto c, d \mapsto d$$

extends to a non-inner central automorphism of H. Let $L = \langle z \rangle$ be of order p^3 acting on H via σ. Then, G is Miller if $N = \langle [a, b] \rangle$, and G is not if $N = \langle [a, d] \rangle$.

Remark 26 The above construction of non-special Miller groups G from special Miller groups H is valid even without hypotheses (i)–(iii) on H. That $\mathrm{Aut}(G) = \mathrm{Autcent}(G)$ can be proved using the same arguments as in [28], which do not rely on hypotheses (i)–(iii). Then, one can apply Theorem 23 to show that $\mathrm{Aut}(G)$ is abelian.

We conclude with the following remarks. In all known examples of Miller p-groups G, $G/Z(G)$ is homocyclic. It will be interesting to know whether this happens in all Millers p-groups. If true, one might expect if $\gamma_2(G)$ is always homocyclic. Again in the known Millers p-groups G one can observe that either G' and $Z(G)$ have same ranks or $G/Z(G)$ and G/G' have same ranks. We wonder whether this is true for all Miller p-groups.

References

1. J.E. Adney, T. Yen, Automorphisms of p-group. Ill. J. Math. **9**, 137–143 (1965)
2. G. Ban, S. Yu, Minimal abelian groups that are not automorphism groups. Arch. Math. **70**, 427–434 (1998)
3. Y. Berkovich, Z. Janko, *Groups of Prime Power Order*, vol. 2 (Walter de Gruyter, New York, 2008)
4. J.N.S. Bidwell, M.J. Curran, D.J. McCaughan, Automorphisms of direct products of finite groups. Arch. Math. **86**, 481–489 (2006)
5. A. Caranti, A Module theoretic approach to abelian automorphism groups. Isr. J. Math. **205**, 235–246 (2015)
6. A. Caranti, Erratum to module-theoretic approach to abelian automorphism groups. Isr. J. Math. **215**, 1025–1026 (2016)
7. M.J. Curran, Semi-direct product groups with abelian automorphism groups. J. Austral. Math. Soc. (Ser. A) **42**, 84–91 (1987)

8. M.J. Curran, Direct products with abelian automorphism groups. Commun. Algebra **35**, 389–397 (2007)
9. R. Dark, C.M. Scoppola, On Camina groups of prime power order. J. Algebra **181**, 787–802 (1996)
10. B. E. Earnley, On finite groups whose group of automorphisms is abelian, Ph.D. Thesis, Wayne State University, 1975
11. R. Faudree, A note on the automorphism group of a p-group. Proc. Am. Math. Soc. **19**, 1379–1382 (1968)
12. R. Faudree, Groups in which each element commutes with its epimorphic images. Proc. Am. Math. Soc. **27**(2), 236–240 (1971)
13. S.P. Glasby, 2-groups with every automorphisms central. J. Austral. Math. Soc. (Ser. A) **41**, 233–236 (1986)
14. D. Gorenstein, *Finite Groups*, 2nd edn. (AMS Chelsea Publication, 1980)
15. H. Heineken, M. Liebeck, The occurrence of finite groups in the automorphism group of nilpotent groups of class 2. Arch. Math. **25**, 8–16 (1974)
16. H. Heineken, Nilpotente Gruppen, deren sm̈atliche Normalteiler charakteristisch sind, Arch. Math. **33** (1979/80), 497 - 503
17. P.V. Hegarty, Minimal abelian automorphism groups of finite groups. Rend. Sem. Mat. Univ. Padova **94**, 121–135 (1995)
18. G. T. Helleloid, *A survey on automorphism groups of finite groups*, arXiv:math/0610294v2 [math.GR], 25 Oct. 2006
19. H. Hilton, *An Introduction to the Theory of Groups of Finite Order* (Oxford at the Clarendon Press, 1908)
20. C. Hopkins, Non-abelian groups whose groups of isomorphisms are abelian. Ann. Math. **29**(1), 508–520 (1927)
21. A. Hughes, Automorphisms of nilpotent groups and supersolvable groups. Proc. Symp. Pure Math. **37**, 205–207 (1980). AMS
22. M.H. Jafari, Elementary abelian p-groups as central automorphism groups. Commun. Algebra **34**, 601–607 (2006)
23. A. Jamali, Some new non-abelian 2-groups with abelian automorphism groups. J. Group Theory **5**, 53–57 (2002)
24. V.K. Jain, M.K. Yadav, On finite p-groups whose automorphisms are all central. Isr. J. Math. **189**, 225–236 (2012)
25. V.K. Jain, P.K. Rai, M.K. Yadav, On finite p-groups with abelian automorphism groups. Int. J. Algebra Comput. **23**, 1063–1077 (2013)
26. D. Jonah, M. Konvisser, Some non-abelian p-groups with abelian automorphism groups. Arch. Math. **26**, 131–133 (1975)
27. T. Karimi, Z.K. Farimani, p-groups with elementary abelian central automorphism groups. World Appl. Program. **1**, 352–354 (2011)
28. R.D. Kitture, M.K. Yadav, Note on Caranti's method of construction of Miller groups. Monatsh. Math. **185**, 87–101 (2018)
29. I. Malinowska, *p-automorphisms of finite p-groups - problems and questions* (Rome, Advances in Group Theory, Aracne Edritice, 2002), pp. 111–127
30. A. Mahalanobis, Diffe-Hellman key exchange protocol and non-Abelian nilpotent groups. Isr. J. Math. **165**, 161–187 (2008)
31. I.D. Macdonald, *The Theory of Groups* (Oxford at the Clarendon Press, 1968)
32. A. Mann, Some questions about p-groups. J. Austral. Math. Soc. (Ser. A) **67**, 356–379 (1999)
33. G.A. Miller, A non-abelian group whose group of isomorphisms is abelian. Messenger Math. **XVIII**, 124–125 (1913-1914)
34. M. Morigi, On the minimal number of generators of finite non-abelian p-groups having an abelian automorphism group. Commun. Algebra **23**, 2045–2065 (1995)
35. M. Morigi, On p-groups with abelian automorphism group. Rend. Sem. Mat. Univ. Padova **92**, 47–58 (1994)
36. R.R. Struik, Some non-abelian 2-groups with abelian automorphism groups. Arch. Math. **39**, 299–302 (1982)

Camina Groups, Camina Pairs, and Generalizations

Mark L. Lewis

MSC [2010] 20D15 · 20C15

1 Introduction

This is a survey of results regarding Camina groups and Camina pairs and related topics. Except where mentioned, all groups are finite. There are no new results in this paper, and we do not include proofs. Our goal is to put in one place all of the main results regarding Camina groups and Camina pairs with one exception. The condition semi-extra special is equivalent to being a nilpotent Camina group of nilpotence class 2. Since we recently wrote an expository article covering the major results regarding semi-extra special groups [47], we are not repeating that content here; instead, we refer readers interested in Camina groups of nilpotence class 2 to consult [47]. We also include many of the generalizations of Camina groups and Camina pairs that have been studied. Finally, we give an extensive list of the problems where these ideas have been applied. We have tried to be as complete as we can; however, we admit the likelihood that we have missed something.

The study of Camina pairs began with [12] where Camina considered two conditions that generalized Frobenius groups and extra special groups. We discuss Camina's conditions in Sect. 4. Macdonald then picked up the study of Camina pairs in [52] and began the study of Camina groups in [52, 53]. We believe that the names Camina group and Camina pair were introduced in [57], but one can see that those names quickly caught on.

M. L. Lewis (✉)
Department of Mathematical Sciences, Kent State University, Kent, OH 44242, USA
e-mail: lewis@math.kent.edu

N. S. N. Sastry and M. K. Yadav (eds.), *Group Theory and Computation*, Indian Statistical Institute Series, https://doi.org/10.1007/978-981-13-2047-7_8

In this paper, we will begin by defining Camina groups and presenting results about general Camina groups. The main result in this section is the characterization of Camina groups by Dark and Scoppola in Theorem 2.4. We then turn to Camina p-groups. As we mentioned above, the exposition of Camina p-groups of nilpotence class 2 can be found in [47]. In this paper, we present known results regarding Camina p-groups of nilpotence class 3.

We then turn to Camina pairs. We begin by defining Camina pairs and mentioning the general results regarding Camina pairs. Among these general results, we include Camina's initial theorem that started the investigation of these topics. We then present the known results regarding the case where (G, K) is a Camina pair such that G/K is a p-group. We then discuss a situation studied by Gagola independently of Camina pairs, but later, it was realized that Gagola's situation was a special case of the Camina pair situation. Next, we present the case where (G, K) is a Camina pair and K is a p-group. Then, we consider the case where (G, K) is a Camina pair and G is a p-group. We follow that by including Camina pairs (G, K) where G is not solvable. We close the section on Camina pairs with the homogeneous induction condition studied by Kuisch and van der Waall.

We then look at generalizations of these ideas. The first is the idea of anti-central elements considered by Ladisch. Next, we present our results about the vanishing-off subgroup of a group. We then present Camina triples. We close this section with a discussion of other generalizations that have been studied.

Finally, we present applications where these ideas have been used. First is Mattarei's amazing examples of groups with identical character tables and different derived lengths. We then discuss weak Cayley tables. After that, we have the solution of Snyder's problem. We close with a long list of other applications. For many of these other applications, we only mention the application, and provide few other details.

We would like to thank the referees for the careful reading of this paper and the helpful suggestions.

2 Camina Groups

The following equivalent conditions can be seen as the basis of Camina groups and the motivation for Camina pairs, and all related objects. The following was proved as Lemma 2.1 in [40], but that result was motivated by Lemma 1 of [12] and Proposition 3.1 of [13].

Lemma 2.1 *Let g be an element of a group G. Then, the following are equivalent:*

1. *The conjugacy class of g is gG'.*
2. $|\mathbf{C}_G(g)| = |G : G'|$.
3. *For every $z \in G'$, there is an element $y \in G$ so that $[g, y] = z$.*
4. $\chi(g) = 0$ *for all nonlinear* $\chi \in \mathrm{Irr}(G)$.

We will call an element g of a group G that satisfies the conditions of Lemma 2.1 a *Camina element*. A nonabelian group G is a *Camina group* if every element in $G \setminus G'$ is a Camina element. There are two key examples of Camina groups (1) any Frobenius group whose Frobenius complement is abelian and (2) any extra special p-group. In particular, the quaternions and the dihedral group of order 8 are Camina groups. As a corollary to Lemma 2.1, we obtain the following equivalent conditions for a group to be a Camina group.

Lemma 2.2 *Let G be a group. Then, the following are equivalent:*

1. *G is a Camina group.*
2. *For every element $g \in G \setminus G'$, the conjugacy class of g in G is gG'.*
3. *For every element $g \in G \setminus G'$, $|\mathbf{C}_G(g)| = |G : G'|$.*
4. *For every element $g \in G \setminus G'$ and for every element $z \in G'$, there is an element $y \in G$ so that $[g, y] = z$.*
5. *Every character $\chi \in \mathrm{Irr}(G)$ vanishes on $G \setminus G'$.*

We make a couple of observations about Camina groups.

Lemma 2.3 *Let G be a Camina group. Then the following are true:*

1. *If N is a normal subgroup of G, then either $N \leq G'$ or $G' \leq N$.*
2. *If $N < G'$, then G/N is also a Camina group.*

The definition of Camina groups was formulated as a common generalization of extra special groups and Frobenius groups. Dark and Scoppola have proved in [16] that there are few other examples:

Theorem 2.4 (Dark and Scoppola) *Let G be a Camina group. Then, one of the following occurs:*

1. *G is a Frobenius group whose Frobenius complement is cyclic.*
2. *G is a Frobenius group whose Frobenius complement is the quaternions.*
3. *G is a p-group for some prime p.*

In particular, the only examples of Camina groups are Frobenius groups whose complements are cyclic, Frobenius groups whose complements are isomorphic to the quaternion group, and Camina groups that are p-groups. We will talk more about Camina p-groups in Sect. 3, and we will see that these groups have many properties in common with extra special groups. Recall that the Frobenius kernel of a Frobenius group must be nilpotent. The following corollary follows immediately from Theorem 2.4.

Corollary 2.5 *If G is a Camina group, then G is solvable.*

At this point, we cannot help but plug our own recent papers [34, 43] where we give a new proof of Theorem 2.4. Our proof relies on the following theorem which is of independent interest. In particular, it has been used by Herzog, Longobardi, and Maj in their work on generalizing the definition of Camina groups to infinite groups (see [27]).

Theorem 2.6 *Let P be a Camina p-group that acts on a nontrivial p'-group Q so that $\mathbf{C}_P(x) \leq P'$ for every $x \in Q \setminus \{1\}$. Then the action of P is Frobenius and P is the quaternions.*

While we do not want to get sidetracked into a discussion regarding infinite Camina groups, we do mention that the situation appears much more complicated in that case. In particular, Herzog, Longobardi, and Maj give examples of infinite Camina groups that are not solvable. Also, infinite Camina groups are studied in [19, 65].

3 Camina p-Groups

Since Frobenius groups are relatively well understood, most of the focus on Camina groups has been on Camina groups that are p-groups. We say that such a group is a *Camina p-group*. The structure of these groups is highly limited. In particular, there is a very tight bound on their nilpotence class. This next theorem gives a flavor of this. It is Theorem 3.1 in [53]. In [55], Mann presents an alternate proof of this next theorem.

Theorem 3.1 (MacDonald) *If G is a Camina 2-group, then G has nilpotence class* 2.

When p is odd, the bound is only slightly less tight. This next result was proved in [16].

Theorem 3.2 (Dark and Scoppola) *If G is a Camina p-group for an odd prime p, then its nilpotence class is either* 2 *or* 3.

In fact, Mann and Scoppola had previously proved the following preliminary result in [57].

Theorem 3.3 (Mann and Scoppola) *Let G be a Camina p-group for an odd prime p and assume that G is Metabelian. Then, G has nilpotence class* 2 *or* 3.

It turns out that Theorem 3.3 is a key step in Dark and Scoppola's proof of Theorem 3.2. In particular, in[16], Dark and Scoppola use Lie rings to show that Theorem 3.2 can be reduced to the case where the group is metabelian, and so they complete the proof of Theorem 3.2 by appealing to Theorem 3.3.

A p-group G is said to be *semi-extraspecial* if for every subgroup N in $Z(G)$ having index p, then G/N is extra special. The following result is proved as Theorem 1.2 in [76].

Theorem 3.4 (Verardi) *Let G be a group. Then, the following are equivalent:*

1. *G is a Camina group of nilpotence class* 2.
2. *G is a semi-extra special p-group for some prime p.*

Thus, the study of Camina p-groups of nilpotent class 2 is precisely the study of semi-extra special groups. We have recently written an expository paper that presents many of the known results regarding these groups [47]. Rather than repeating that work here, we refer the reader to that paper.

One particular family of semi-extra special groups should be highlighted. We define the *Heisenberg group* of degree n for the prime p to be a Sylow p-subgroup of $\mathrm{GL}_3(p^n)$. That is, this is the group of 3 by 3 upper triangular matrices with 1's on the diagonal and entries from the field of order p^n. Note that it has order p^{3n}.

3.1 Camina p-Groups of Nilpotence Class 3

We now turn to Camina groups of nilpotence class 3. Recall from Theorem 3.1 if G is a Camina 2-group, then G has nilpotent class 2. Therefore, when discussing a Camina p-group of nilpotence class 3, we necessarily know that p is an odd prime.

We begin with a theorem (Theorem 3.5) that lists most of the known results about these groups.

Theorem 3.5 *Let G be a Camina p-group of nilpotence class 3. Let $C = C_G(G')$. Then, the following are true:*

1. *G/G', G'/G_3, and G_3 are elementary abelian where $G_3 = [G', G]$.*
2. *$G_3 = \mathbf{Z}(G)$ and $G'/G_3 = \mathbf{Z}(G/G_3)$.*
3. *$|G : G'| = p^{2n}$, $|G' : G_3| = p^n$, and $|G_3| < p^{3n/2}$ for some even integer n.*
4. *G/G_3 and G' have exponent p.*
5. *Either $|C : G'| = |G' : G_3| = p^n$ or $C = G'$.*
6. *C/G_3 is an elementary abelian p-group.*
7. *If $|G_3| = p$, then $|C : G'| = p^n$.*
8. *If $|G_3| \geq p^n$, then G/G_3 is isomorphic to the Heisenberg group of degree n for the prime p.*
9. *Every nonlinear character in $\mathrm{Irr}(G/G_3)$ is fully ramified with respect to G/G' and every character in $\mathrm{Irr}(G \mid G_3)$ is fully ramified with respect to G/Z.*
10. $\mathrm{cd}(G) = \{1, p^n, p^{3n/2}\}$.

We now list the references for the results included in Theorem 3.5. Conclusions (1), (2), and the first two parts of (3) were proved by MacDonald as Corollary 2.3, Theorem 5.2 (ii), and Corollary 5.3 in [52]; the last part of (3) was proved by Dark and Scoppola in [16]; and Conclusion (4) was proved by Mann in [55]. Conclusion (5) is a corollary of Lemma 5.1 of [52] with some of the arguments used in the proof of Theorem 5.2 of [52]. (A proof of Conclusion (5) in a more general context can be found as Lemma 4.11 of [41].) Conclusion (6) is a consequence of Theorem 1.3 (iii) and (iv) of [57]. (An independent proof of Conclusion (6) can be found as Proposition 4 of [34].) Conclusion (7) was proved by Macdonald in [52]; Conclusion (8) was proved by Mann and Scoppola in [57]; and Conclusions (9) and (10) were proved by the author along with Moretó and Wolf in [48].

Recall from Lemma 2.3 that if G is a Camina group and $N < G'$ is a normal subgroup of G, then G/N is also a Camina group. In particular, if G is a Camina group of nilpotence class 3, then G/G_3 must be a Camina group of nilpotence class 2. This fact is used in proving (1), (2), and (3) of Theorem 3.5.

In Sect. 6 of [52], MacDonald constructs for every odd prime p and every even positive integer n, a Camina p-group G such that $|G| = p^{3n+1}$, G has nilpotence class 3, and $|G_3| = p$. MacDonald also constructs a Camina group G of nilpotence class 3 and order 3^8 where $|G : G'| = 3^4$ and $|G' : G_3| = |G_3| = 3^2$ in [53]. In Sect. 4 of [16], Dark and Scoppola construct for any odd prime p and every positive even integer n, a Camina p-group G of order p^{4n} having nilpotence class 3 and satisfying $|G_3| = p^n$ where $|G : G'| = |G_3|^2$.

One note on conclusion (3) of Theorem 3.5. MacDonald sketches out a proof of the stronger assertion that $|G_3| \leq p^n$ in Lemma 2.1 of [53], but Dark and Scoppola point out in Sect. 1 in [16] that there is a problem with Macdonald's sketched proof. Dark and Scoppola prove in [16] $|G_3| < p^{3n/2}$. At this time, there are no known Camina groups G of nilpotence class 3 with $p^n < |G_3| < p^{3n/2}$, so it is possible that Macdonald's assertion is true. In particular, when G satisfies the equation $|G_3| \leq p^n$, we say that G satisfies the *Macdonald Bound*. At this time, we have the following open question:

Open Question: Does the MacDonald bound hold for all Camina groups of nilpotence class 3? That is, does there exists a Camina group G of nilpotence class 3 with $|G : G'| = p^{2n}$ for the prime p and positive integer n with $p^n < |G_3| < p^{3n/2}$?

Notice that the smallest value of n for which there might possibly be a Camina group G of nilpotence class 3 that violates the MacDonald bound is $n = 4$, and in this case, such a group G would have $|G : G'| = p^8$, $|G' : G_3| = p^4$, and $|G_3| = p^5$; so $|G| = p^{17}$.

In our paper, [46], we consider bounds on $|G_3|$ when G is a Camina group of nilpotence class 3. We are not able to approach the MacDonald bound, but we do obtain some interesting results in some special cases. We note that $|G' : G_3|$ is a square, so $|G' : G_3|^{1/2}$ is necessarily an integer. The following is Theorem 1 of [46].

Theorem 3.6 *If G is a Camina p-group of nilpotence class* 3 *and $G' < C_G(G')$, then $|G_3| \leq |G' : G_3|^{1/2}$.*

In the next theorem, we remove the hypothesis that $G' < C_G(G')$. However, we do add a hypothesis on the number of abelian subgroups of the quotient G/G_3. This is Theorem 2 of [46].

Theorem 3.7 *Let G be a Camina p-group of nilpotence class* 3, *and let $H = G/G_3$.*

1. *If H has one or two abelian subgroups of order $|H : H'|$, then $C_G(G') > G'$.*
2. *If H has $p^a + 1$ abelian subgroups of order $|H : H'|$ for the positive integer a, then either $C_G(G') > G'$ or $|G_3| \leq p^{2a}$.*

Note that, Theorem 3.5 (8) shows that if G is a Camina group of nilpotence class 3 with $|G_3| \geq |G : G'|^{1/2}$, then G/G_3 is isomorphic to the Heisenberg group. On the other hand, if $|G_3| < |G : G'|^{1/2}$, then we have much less information regarding G/G_3. We do know that G/G_3 must be a Camina group and it must have nilpotence class 2; so by Theorem 3.4 we know that G/G_3 must be semi-extra special. Since $G_3 > 1$, we also know that there will a subgroup $N < G_3$ so that $|G_3 : N| = p$. Letting $C/N = C_{G/N}(G'/N)$ and applying Theorem 3.5 (7), we see that C/G_3 will be abelian. This implies that G/G_3 has an abelian subgroup of order $|G : G'|$. Not all semi-extra special groups have an abelian subgroup of this order. In [46], we present two nonisomorphic semi-extra special groups that are not isomorphic to the Heisenberg group, but it do occur as the quotients of Camina groups of nilpotence class 3. This leads us to ask the following question.

Open Question: Which semi-extra special groups can occur as quotients of Camina groups of nilpotence class 3?

We close by mentioning the following result proved by Mann in Theorem 3 in [55].

Theorem 3.8 (Mann) *If G is a Camina 3-group of nilpotence class 3, then G has exponent 9.*

In general, if G is a Camina p-group of nilpotence class 3, then Theorem 3.5 (1) implies that the exponent of G is at most p^2. We note the groups constructed by Dark and Scoppola mentioned above have exponent p when $p \geq 5$, so $p = 3$ is an exception.

4 Camina Pairs

Camina pairs arise by generalizing the conditions in Lemma 2.2 where we replace G' with some arbitrary normal subgroup K in G. In particular, let G be a group and $1 < K$ be a proper normal subgroup of G. Then, (G, K) is a *Camina pair* if for every element $g \in G \setminus K$, then g is conjugate to every element of gK. Notice that, this says that the conjugacy class of g is a union of cosets of K in G.

Now, when $K = G'$, this is the precisely that condition that we used to define Camina groups. Hence, a group G is a Camina group if and only if (G, G') is a Camina pair. In many places, the term *Camina kernel* is used for the subgroup K. That is. K is a Camina kernel for G if and only if (G, K) is a Camina pair.

Since Lemma 2.2 has a number of equivalent conditions, it should not be surprising that there are a large number of equivalent conditions for Camina pairs. In Lemma 4.1, we list a number of these equivalent conditions. If K is a normal subgroup of G, then we write $\mathrm{Irr}(G \mid K)$ for the set of irreducible characters of G whose kernels do not contain K.

Lemma 4.1 *Let $1 < K$ be a proper normal subgroup of a group G. Then, the following are equivalent:*

1. *(G, K) is a Camina pair.*

2. *If $x \in G \setminus K$, then $|\mathbf{C}_G(x)| = |\mathbf{C}_{G/K}(xK)|$.*
3. *If xK and yK are conjugate in G/K and nontrivial, then x is conjugate to y in G.*
4. *For all $x \in G \setminus K$ and $z \in K$, then there exists an element $y \in G$ so that $[x, y] = z$.*
5. *If $C_1 = \{1\}, \ldots, C_m$ are the conjugacy classes of G contained in K and $C_{m+1}, \ldots, C_n$ are the conjugacy classes of G outside K, then $C_i C_j = C_j$ for $1 \leq i \leq m$ and $m + 1 \leq j \leq n$.*
6. *For all $\chi \in \mathrm{Irr}(G \mid K)$, then χ vanishes on $G \setminus K$.*
7. *Every nonprincipal character in $\mathrm{Irr}(K)$ induces homogeneously to G.*

We now list the references for the results included in Lemma 4.1. That Condition (2) follows from Conditions (3) and (4) is proved in Lemma 1 of [12] Conditions (2), (3), (4), and (6) were proved equivalent to the Camina condition in Proposition 3.1 of [13]. Condition (5) was proved equivalent to Condition (6) in [71]. Condition (7) was proved equivalent to the Camina condition in Lemma 2.1 of [37].

We saw above that G is a Camina group if and only if (G, G') is a Camina pair. When G is a Camina group of nilpotence class 3, we can say more. Macdonald proved the following as Theorem 5.2 (i) of [52].

Theorem 4.2 (Macdonald) *Let G be a Camina p-group of nilpotence class* 3. *Then (G, G_3) is a Camina pair.*

Some other facts regarding Camina pairs are reasonably obvious.

Lemma 4.3 *Let (G, K) be a Camina pair. Then, the following are true:*

1. *If $N < K$ is normal in G, then $(G/N, K/N)$ is a Camina pair.*
2. *$Z(G) \leq K \leq G'$.*

Recall that if G is a Frobenius group where a Frobenius complement is either cyclic or quaternion, then G is a Camina group. On the other hand, it is not difficult to see that if G is any Frobenius group with Frobenius kernel K, then (G, K) is a Camina pair. Thus, generalizing from Camina groups to Camina pairs, we capture all Frobenius groups. As we mentioned earlier, K is sometimes called a Camina kernel when (G, K) is a Camina pair, and in this sense, we can think of the Camina kernel as being a generalization of the Frobenius kernel. In Theorem 2 of [12], Camina proved the first structural result regarding Camina pairs.

Theorem 4.4 (Camina) *If (G, K) is a Camina pair, then either G is Frobenius group with Frobenius kernel K or one of G/K or K is a p-group for some prime p.*

We will use Theorem 4.4 to split Camina pairs into different categories. In Proposition 3.2 of [13], Chillag and Macdonald find a way to distinguish the Camina pairs that are Frobenius groups. Recall that a group G splits over a normal subgroup K if there is a subgroup H that complements K. That is, $G = KH$ and $K \cap H = 1$.

Theorem 4.5 (Chillag and Macdonald) *If (G, K) is a Camina pair, then G is a Frobenius group with Frobenius kernel K if and only if G splits over K.*

This gives the immediate corollary which was actually first proved by Camina in [12].

Corollary 4.6 (Camina) *If (G, K) be a Camina pair, then G is a Frobenius group with Frobenius kernel K if and only if $(|G : K|, |K|) = 1$.*

While we are discussing general results for Camina pairs, we mention a couple of other lemmas that appear in [13] that seem useful. Lemma 4.7 shows that Camina pairs can be put together to obtain another Camina pair.

Lemma 4.7 *If (G, K) and $(G/K, N/K)$ are Camina pairs, then (G, N) is a Camina pair.*

Lemma 4.8 shows that certain subgroups of Camina pairs yield new Camina pairs.

Lemma 4.8 *If (G, K) is a Camina pair and $G = HK$, then $(H, H \cap K)$ is a Camina pair.*

Lemma 4.9 was proved in [13]. This yields information about a Sylow subgroup of a Camina pair.

Lemma 4.9 *Suppose (G, K) is a Camina pair where either G/K or K is a p-group and G is not a Frobenius group with Frobenius kernel K. If P is a Sylow p-subgroup of G, then $Z(P) \leq K$ and in particular, P is not abelian.*

The following generalizes a result of Gagola (see Corollary 2.3 of [20]). Lemma 4.10 appears as Proposition 5.1 in [37].

Lemma 4.10 (Kuisch, van der Waall) *Suppose (G, K) is a Camina pair. If π is the set of primes that divide $|K|$ and $\nu \in \mathrm{Irr}(K)$ is a nonprincipal character, then the stabilizer in G of ν is a π-group. In particular, if K is a p-group for some prime p, then the stabilizer in G of ν is a p-group.*

We do want to mention the result which started the study of Camina pairs. This study began with the following observation of Camina that was Theorem 1 of [12]. Camina's motivation came from the fact that both Frobenius groups and extra special groups satisfy both of these properties.

Theorem 4.11 (Camina) *Let G be a group, and let N be a nontrivial proper normal subgroup of G. Then, the following two conditions are equivalent:*

1. *If $x \in G \setminus N$, then x is conjugate to every element in xN.*
2. *There exist irreducible characters $\chi_1, \ldots, \chi_n$ of G so that each χ_i vanishes on $G \setminus N$ and there exist positive integers $a_1, \ldots, a_n$ so that $\sum_{i=1}^{n} a_i \chi_i$ is constant on $N^\#$.*

Notice that, Condition (1) of [12] is the definition of Camina pairs, so Camina pairs can be characterized by Condition (2). We define $\mathrm{Irr}(G \mid N)$ to be the set of the irreducible characters of G whose kernels do not contain N. It is not difficult to show in Condition (2) that $\mathrm{Irr}(G \mid N) = \{\chi_1, \ldots, \chi_n\}$.

4.1 G/K a p-Group

In the next several subsections, we will look at Camina pairs in more detail. Recall from Theorem 4.4 that if (G, K) is a Camina pair where G is not a Frobenius group with Frobenius kernel K, then either G/K or K is a p-group. In this first section, we consider Camina pairs (G, K) where G/K is a p-group for some prime p. In Theorem 2.1 of [33], Isaacs proved the following theorem which strengthens Theorem 4.4.

Theorem 4.12 (Isaacs) *Suppose (G, K) is a Camina pair such that G/K is nilpotent. Then, either*

1. *G is a Frobenius group with Frobenius kernel K, or*
2. *G/K is a p-group for some prime p, G has a normal p-complement M, and $\mathbf{C}_G(m) \le K$ for all $m \in M \setminus \{1\}$.*

Since Frobenius groups have normal p-complements for every prime p, the following result is a corollary to the theorem labeled as Theorem C of [33]. This result was proved independently in [14].

Theorem 4.13 (Isaacs) *If (G, K) is a Camina pair and G/K is a p-group for some prime p, then G has a normal p-complement.*

We now look for examples of Camina pairs (G, K) where G/K is a p-group. Obviously, when G is a Frobenius group and K is a Frobenius kernel whose Frobenius complement is a p-group, (G, K) will be such an example. Another example is when (G, G') where G is a Camina p-group. We have also seen that if G is a Frobenius group whose Frobenius complement is isomorphic to the quaternions, then (G, G') is a Camina pair where G/G' has order 4. At this time, no other Camina pairs (G, K) where G/K is a p-group are known.

The question of the existence of other examples has been considered in [13, 14]. In [13], they proved that there is no other example whose Sylow p-subgroup has nilpotence class at most 2. In [14], they take this further.

Theorem 4.14 (Chillag, Mann, Scoppola) *Suppose (G, K) is a Camina pair such that G/K is a p-group and K is not a p-group. If P is a Sylow p-subgroup of G, then P does not satisfy any of the following conditions (among others): (1) P is of maximal class, (2) the nilpotence class of P is at most $p + 1$, or (3) $K \cap P = Z(P)$.*

In the situation of Theorem 4.14, they do not prove, but they state that they have a proof that the case where $p = 2$ and P has nilpotence class 4 cannot occur.

In Proposition 11 of [55], Mann proves the following curious necessary condition:

Theorem 4.15 (Mann) *Let (G, K) be a Camina pair with G/K a p-group and G not a p-group for some odd prime p. Then any two maximal cyclic subgroups of G/K that have a nontrivial intersection are conjugate. The same conclusion holds for $p = 2$, provided the intersection has size at least* 4.

Mann mentions that groups of exponent p, cyclic groups, and dihedral groups are obvious examples of groups satisfying the conjugacy condition of Theorem 4.15. Mann notes one group of order 81 that satisfies this condition and other than the obvious groups, it is the only group of order p^4 that satisfies this condition. He mentions that there exist examples of orders p^5 and p^6. he stated that he will return to these groups in a later paper, but to our knowledge, no one has published any further results regarding these groups.

Given all of this evidence, the following seems like a reasonable conjecture:

Conjecture *Let (G, K) be a Camina pair and assume that G/K is a p-group for some prime p. Then, one of the following occurs: G is a Frobenius group with Frobenius kernel K and Frobenius complement that is a p-group, G is a Camina p-group with $K = G'$, or G is a Frobenius group whose Frobenius complement is the quaternions and $K = G'$.*

4.2 Gagola Characters

For the remaining subsections regarding Camina pairs, we consider Camina pairs where the normal subgroup is a p-group. Before looking at these Camina pairs in detail, we make a detour into a condition that was studied independently. In [20], Gagola studied the following condition. Suppose a group G has an irreducible character χ that vanishes on all but two conjugacy classes of G. He proved the following as Lemma 2.1 in [20].

Theorem 4.16 (Gagola) *Let G be a group with $|G| > 2$ that has an irreducible character χ that does not vanish on exactly two conjugacy classes of G. Then χ is unique, and it is the unique faithful irreducible character of G. Furthermore, G contains a unique minimal normal subgroup N. The subgroup N is an elementary abelian p-group for some prime p. The character χ vanishes on $G \setminus N$ and is nonzero on N. Finally, the action of G by conjugation on N is transitive on the nonidentity elements of N.*

If G with $|G| > 2$ has an irreducible character χ so that χ does not vanish on exactly two conjugacy classes of G, then we will say that χ is a *Gagola character* and G is a *Gagola group*. If G is a Gagola group, then we apply Theorem 4.16 to see that G has a unique minimal normal subgroup N. It is not difficult to see that $\mathrm{Irr}(G \mid N) = \{\chi\}$ where χ is the Gagola character, and we know that χ vanishes on $G \setminus N$. By Lemma 4.1, we see that (G, N) is a Camina pair where N is a p-group, so all of the results of Sect. 4.3 will apply to Gagola groups. In particular, the groups Gagola studied are a special case of Camina pairs.

It is not difficult to see that extra special 2-groups and doubly transitive Frobenius groups will be examples of Gagola groups. Furthermore, if G is a p-group and a Gagola group, then $|Z(G)| = 2$, and thus, G is a 2-group. In particular, if G is a p-group, then G is a Gagola group if and only if $(G, Z(G))$ is a Camina pair and

$|Z(G)| = 2$. If G is a Frobenius group, then G is a Gagola group if and only if G is a doubly transitive Frobenius group.

Prof. Gagola's work predates most of the results on Camina pairs. At least, a couple of the results that Gagola first proved for Gagola groups were later proved for Camina pairs. For example, Gagola proved the following as Theorem 2.5 in [20].

Theorem 4.17 (Gagola) *Let G be a Gagola group and write N for the unique minimal normal subgroup of G. Suppose p is the prime so that N is a p-group. If P is a Sylow p-subgroup of G, then $Z(P) \leq N$, $Z(P) = N$ if and only if P is normal in G, and N is a term of the upper central series for G.*

The following theorem was proved as Theorem 6.2 of [20] gives more structure on the p-subgroups of a Gagola group.

Theorem 4.18 (Gagola) *Let G be a Gagola group, and let N be the unique minimal normal subgroup which we know is an elementary abelian p-group for some prime p. Then, $N = \mathbf{O}_p(G)$ if and only if G is a doubly transitive Frobenius group or $|G| = 2$.*

In Theorem 6.3 of [20], Gagola shows that every p-group shows up as a subgroup of a Sylow p-subgroup of G/N where G is a Gagola group and N is the unique minimal normal subgroup and N is a p-group.

Theorem 4.19 (Gagola) *Let p be a prime, let Q be any p-group, and let a be a positive integer. Then, there exists a Gagola group G with a normal Sylow p-subgroup P and a cyclic group H of order $p^a - 1$ so that $G = PH$, $Z(P)H$ is a doubly transitive Frobenius group of order $p^a(p^a - 1)$ and Q is isomorphic to a subgroup of $P/Z(P)$. Notice that in this case, $Z(P)$ is the unique minimal normal subgroup of G and $|Z(P)| = p^a$.*

Observe that Theorem 4.19 implies that Gagola groups are going to be plentiful. Gagola also presents two examples of Gagola groups where the Sylow p-subgroup is not normal when the unique minimal normal subgroup is a p-group. The construction of these examples appears on p. 383 and 384 of [20]. They are also described in detail on p. 274–275 in [14]. Both examples presented are $\{2, 3\}$-groups. One has a unique minimal normal 2-subgroup whose Sylow 2-subgroup is not normal. The other has a unique minimal normal 3-subgroup whose Sylow 3-subgroup is not normal.

In Lemma 3.1 of [45], we found another characterization of Gagola groups.

Lemma 4.20 *Let G have a minimal normal subgroup N that is an elementary abelian p-group for some prime p. Then, $\chi \in \mathrm{Irr}(G)$ is a Gagola character for G if and only if $\chi(1) = (|G| - |G : N|)^{1/2}$.*

Recently, Wang, Chang, and Jin generalized the idea of Gagola characters in [77]. In their paper, they consider a group G and a character χ with the property that number of conjugacy classes that χ does not vanish on is exactly one more than number of conjugates of χ under the action on the characters of G by the Galois group of the field obtained by extending the rational field by the values of χ over the rational field. They show that these groups also yield a Camina pair, and they prove number of interesting results regarding these groups.

4.3 K is a p-Group

We now study the general case of a Camina pair (G, K) where K is a p-group. Recall that we do not know of any Camina pair (G, K) where G/K is a p-group, but G is neither a Frobenius group nor a p-group. The situation when K is a p-group is quite different. We know of many examples where (G, K) is a Camina pair such K is a p-group and G is neither a Frobenius group nor a p-group. In particular, the examples due to Gagola include those that are presented in Theorem 4.18 and the two examples mentioned in the penultimate paragraph of Sect. 4.2. We will later mention other examples of these pairs.

We now gather results regarding Camina pairs (G, K) where K is a p-group. Let P be a Sylow p-subgroup of G. We know that $K \leq P$. By Lemma 4.6, we know that G is a Frobenius group if and only if $P = K$. Given a group G, a *central series* is a series of subgroups $1 = N_0 \leq N_1 \leq \cdots \leq N_n$ so that $N_{i+1}/N_i \leq Z(G/N_i)$ for all i so that $0 \leq i \leq n - 1$. The group G is nilpotent if and only if G has a central series with $N_n = G$. When G is nilpotent, we only consider central series that end with G. We now present Proposition 1.1 of [14] which shows that K will be a term in every central series for P. Notice that, this generalizes the last conclusion of Theorem 4.17.

Lemma 4.21 *Let (G, K) be a Camina pair with K a p-group and let P be a Sylow p-subgroup of G. Then, K appears as a term in every central series of P.*

The forward direction of this next theorem is presented as Lemma 4.2 in [13]. The converse is proved as Theorem 1.1 in [75].

Theorem 4.22 (Kuisch, van der Waall) *Let (G, K) be a Camina pair where K is a p-group for some prime p. Let P be a Sylow p-subgroup of G. Then, P is normal in G if and only if (P, K) is a Camina pair.*

The forward direction of Theorem 4.22 can be generalized. The following is Theorem 5.4 of [37].

Theorem 4.23 (Kuisch, van der Waall) *Let (G, K) be a Camina pair where K is a p-group for some prime p. If M is a normal subgroup of G and $K < M$ and p does not divide $|G : M|$, then (M, K) is a Camina pair.*

We know that if G is a Gagola group, then G has a normal p-subgroup for some prime p so that (G, N) is a Camina pair. Applying Theorem 4.22, we see that if G is a Gagola group and P is a normal Sylow p-subgroup of G, then (P, N) will be a Camina pair. With this in mind, we may appeal to Theorem 4.19 to obtain the following.

Corollary 4.24 *Let p be a prime, let Q be a p-group, and let $a > 0$ be an integer. Then, there exists a p-group P so that $(P, Z(P))$ is a Camina pair, $|Z(P)| = p^a$, and Q is isomorphic to a subgroup of $P/Z(P)$.*

A group is said to be p-closed if its Sylow p-subgroup is normal. At one point, people asked whether (G, K) is a Camina pair where K is a p-group implies that G is p-closed. (See p. 112 of [13].) It was then realized that the groups constructed by Gagola in [20] are Camina pairs. As we saw in Sect. 4.2, two of the examples presented by Gagola give Camina pairs (G, K) with K a p-group where G is not p-closed. One example has $p = 2$ and the other example has $p = 3$. At that point, it was conjectured that Camina pairs of the form (G, K) where K is a p-group and p is not 2 or 3 have that G is p-closed. (See the comments in the Example on p. 274 of [14] and on p. 401 of [36].) However, we prove the following theorem as Theorem 1.1 in [45] which disproves this conjecture.

Theorem 4.25 *Let p be a prime. Then, there exists a solvable group G with a normal p-subgroup K so that (G, K) is a Camina pair and G is not p-closed. In fact, G can be chosen to be a Gagola group.*

In other words, there are examples where G is not p-closed for every prime p. Kuisch proves the following as Theorem A of [36]. A *normal p-series* for a group G is a series of normal subgroups $1 = K_0 \le L_1 \le K_1 \le L_2 \le K_2 \le \cdots \le L_n \le K_n = G$ so that p does not divide $|L_i : K_{i-1}|$ and K_i/L_i is a p-group for all integers i with $1 \le i \le N$. We know that G is p-solvable if and only if G has a p-series. Assuming G is p-solvable, the *p-length* of G is the smallest n so that G has a p-series with $K_n = G$. In particular, G has *p-length one* if there exists normal subgroups $L \le K$ in G so that p does not divide $|L|$ and $|G : K|$ and K/L is a p-group. Notice that if G is p-closed, then G has p-length one, so this next theorem is really about the situation when G is not p-closed.

Theorem 4.26 (Kuisch) *Let (G, K) be a Camina pair where G is solvable and K is a p-group. Then, $\mathbf{O}_{p'}(G) = 1$ and the p-length of G is at most 2.*

In Theorem B of [36], Kuisch gets more detailed information about the structure of G under the hypotheses of Theorem 4.26 when G is not p-closed, however, these conditions are complicated and not particularly enlightening so we do not include them here. Under some additional hypotheses, one can obtain the conclusion that G is p-closed. The following is a combination of Theorem 4.1 in [13] and Corollary 1.1 of [36].

Theorem 4.27 *Let (G, K) be a Camina pair with K a p-group for some prime p. Let P be a Sylow p-subgroup of G. Assume one of the following conditions holds:*

1. *G is p-solvable and P/K is abelian.*
2. *G is p-solvable and P has nilpotence class at most 2.*
3. *K is cyclic.*
4. *G is solvable and P has no factor group isomorphic to $Z_p \wr Z_p$.*
5. *G is solvable and $|K| < p^p$.*
6. *G is solvable and P' is normal in G.*
7. *G is solvable and P has maximal class.*

Then, P is normal in G. Furthermore, G is a semi-direct product HP where H is a Hall p-complement of G, and if H is nontrivial, then HK is a Frobenius group with Frobenius kernel K.

The following result gives a necessary and sufficient condition that includes the Sylow p-subgroup being normal. When G is a group and χ is a character of G, the *vanishing-off subgroup* of χ is the subgroup $V(\chi) = \langle g \in g \mid \chi(g) \neq 0 \rangle$. This next result appears as Lemma 5.6 and Proposition 5.7 of [37].

Theorem 4.28 (Kuisch, van der Waall) *Let (G, K) be a Camina pair with K a p-group for some prime p. Let P be a Sylow p-subgroup of G. Then, $K = Z(P)$ if and only if P is normal in G and $K = V(\theta)$ for all $\theta \in \mathrm{Irr}(P \mid N)$.*

This next lemma shows that p'-subgroups are Frobenius complements when (G, K) is a Camina pair having K as a p-group. This appears as Lemma 4.3 in [13].

Lemma 4.29 *Let (G, K) be a Camina pair where K is a p-group for some prime p. If T is any p'-subgroup of G, then TK is a Frobenius group with Frobenius kernel K and Frobenius complement T. In particular, every Sylow subgroup of T is either cyclic or generalized quaternion.*

In this case, there are a number of examples that are not Frobenius groups. In particular, the Gagola groups mentioned Theorem 4.19 and in the paragraph that followed that theorem are such examples. We now present some examples that are not Gagola groups and not Frobenius groups. In this section, we will focus on examples where G is not a p-group. We will save the case where G is a p-group until later. The examples we present in this section all arise from the following observation of Chillag and Macdonald in Lemma 2.2 of [13].

Lemma 4.30 *Let $G = PT$ where P and T are subgroups of G with $Z(P) = K$ where K is a normal subgroup of G and $K \leq T$. If (P, K) is a Camina pair and T is a Frobenius group with Frobenius kernel K, then (G, K) is a Camina pair.*

First, suppose G is a Frobenius group with Frobenius kernel P, and suppose P is a p-group where $(P, Z(P))$ is a Camina pair. This Lemma shows that $(G, Z(P))$ is a Camina pair since if H is a Frobenius complement for G, then $Z(P)H$ will be a Frobenius group. In Example 2 of [13], they showed that there exists a Frobenius group whose Frobenius kernel is an extra special group of order 7^3 and exponent 7 and whose Frobenius complement is cyclic of order 3. This example is far from unique, and it is easy to find other primes and more complicated examples.

There also are examples where G is not a Frobenius group. Let C be a cyclic group of order p^n where p is an odd prime, and let A be the automorphism group of C. Take G to be the semi-direct product of A acting on C. Notice that, A is cyclic of order $p^{n-1}(p-1)$, so G is metacyclic of order $p^{2n-1}(p-1)$. If one takes P to be the Sylow p-subgroup of G, it is not difficult to see that $(P, Z(P))$ is a Camina pair, and if H is a subgroup of order $p - 1$, then $Z(P)H$ is a Frobenius group. This example appeared as Example 3 of [13].

In Example 4 of [13], Chillag and Macdonald showed that the quaternions act on a Camina group of nilpotence class 2 of order 3^6 whose center has order 3^2 where the action on the center is a Frobenius action. (The whole action is not Frobenius.)

4.4 G is a p-Group

In the previous section, we considered the Camina pairs (G, K) where either G/K is a p-group or K is a p-group, but we did not assume that G was a p-group. In fact, we considered a number of results and examples where G is not a p-group. In this section, we now suppose that (G, K) is a Camina pair where G is a p-group.

These groups have been studied in [52] by Macdonald. The first result is Theorem 7.1 of [52].

Theorem 4.31 (Macdonald) *If (G, K) is a Camina pair with G a p-group for some prime p, then $|G : K|$ is an even power of p.*

We will use the following notations. A series of normal subgroup $N_0 < N_1 < \cdots < N_r$ in G is called a central series if $N_i/N_{i-1} \leq Z(G/N_{i-1})$ for $i = 1, \ldots, r$. If $N_0 = 1$ and N_r, then we say that G is nilpotent. There are two important central series. The first is the lower central series, which is given by $G_1 = G$ and $G_i = [G_{i-1}, G]$ for $i > 1$. Also, we have the upper central series which is given by $Z_0(G) = 1$ and $Z_1(G) = \mathbf{Z}(G)$ and $Z_i(G)/Z_{i-1}(G) = \mathbf{Z}(G/Z_{i-1}(G))$ for $i > 1$. A group G has nilpotence class c if and only if $Z_c(G) = G$ and $Z_{c-1}(G) < G$, and this is true if and only if $G_{c+1} = 1$ and $G_c > 1$. One key lemma that he proved is the following as Lemma 2.1 of [52].

Lemma 4.32 *If (G, K) is a Camina pair where G is a p-group, then K is a term in both the upper and lower central series for G.*

In particular, if the nilpotence class of G is c, then there is a positive integer r so that $K = G_r = Z_{c-r+1}(G)$.

In particular, this proves that if (G, K) is a Camina pair where G is a p-group, then K is a term in both the upper central series and the lower central series of G. Macdonald also gets information when $K = \mathbf{Z}(G)$. This is Theorem 2.2 of [52].

Theorem 4.33 (Macdonald) *Let (G, K) be a Camina pair where G is nilpotent of class c. If $K = Z_r(G)$ where $1 \leq r < c$, then $Z_i(G)/Z_{i-1}(G)$ has exponent p for $r \leq i \leq c$.*

Macdonald then obtains the following corollary as Corollary 2.3 of [52].

Corollary 4.34 (Macdonald) *Let (G, K) be a Camina pair where G is nilpotent of class c. If $K = G_r$ where $2 \leq r \leq c$, then G_i/G_{i+1} has exponent p for $r - 1 \leq i \leq c$.*

We note that we really do need $r - 1 \leq i$ in this previous corollary. Take G to be the group of order 32 given by

$$\langle a, b, c, d, e \mid a^2 = d, b^2 = c^2 = d^2 = e^2 = 1, b^a = b * c, c^a = c * e, d^b = d * e \rangle.$$

This is SmallGroup(32, 6) from the small group library in Magma. It is not difficult to see that G has nilpotence class 3. Also, (G, G_3) is a Camina pair where $G_3 = Z(G)$. However, G/G' has exponent 4, not 2. Since $G/Z_2(G)$ has exponent 2, this implies that $G' < Z_2(G)$.

From Corollary 4.24, we see that Camina pairs of the form $(G, Z(G))$ where G is a p-group are plentiful. The only known examples of Camina pairs (G, N) where G is a p-group and $N \neq Z(G)$ are (G, G') where G is a Camina p-group of nilpotence class 3. Hence, Macdonald has the following as Conjecture 2 in [53]:

Conjecture *If (G, N) is a Camina pair with G a p-group having nilpotence class c, then either $N = G_{c-1}$ or $N = G_c$.*

Notice that if this conjecture is true, then it would imply that either $N = Z(G)$ or $N = Z_2(G)$. Macdonald also has the following as Conjecture 1 in [53]. This conjecture is based on the fact that it is true for Camina groups of nilpotence class 3.

Conjecture *If (G, G_i) is a Camina pair, then (G, G_{i+1}) is also a Camina pair.*

At this point, there has not been much study of Camina pairs (G, K) where G is a p-group beyond Macdonald's work. But we do mention one result that we have proved in this situation. In [42], we give an easy argument showing that if $(G, Z(G))$ is a Camina pair, then G must be a p-group. In Theorem 3 of [42], we prove the following:

Theorem 4.35 *If $(G, Z(G))$ is a Camina pair, then $|Z(G)| < |G : Z(G)|^{3/4}$.*

Under additional hypotheses, we can improve this bound to a square root, and we know of no examples where the square root is exceeded. Thus, it makes sense to ask:

Open Question: If $(G, Z(G))$ is a Camina pair, is it true that $|Z(G)| \leq |G : Z(G)|^{1/2}$?

A possible generalization of Theorem 4.35 can be found in [54].

4.5 Nonsolvable Camina Pairs

Observe that Camina groups are necessarily solvable. On the other hand, we know that Frobenius groups give rise to Camina pairs. Since there exist nonsolvable Frobenius groups, there must exist nonsolvable Camina pairs. The obvious question becomes: Do there exist nonsolvable Camina pairs that are not Frobenius groups? This question was answered positively in [5]. In that paper, they present an example of a group G with $|G| = 2^3 \cdot 3 \cdot 5^6 =$ with a normal subgroup N so that $|N| = 5^2$, (G, N) is a

Camina pair, $G/O_5(G) \cong \mathrm{SL}(2,5)$, and G is not a Frobenius group. We note that G is a Gagola group; so this is also the first nonsolvable Gagola group that is not a Frobenius. Obviously, G is not 5-closed, so this was the first example of both a Camina pair and a Gagola group that is not 5-closed. As explained in [5], the construction of this group does not generalize to other primes.

In Theorem 1.20 of [24], we present examples of p-closed Camina pairs (G, K) where $|K| = p^2$ and $G/\mathbf{O}_p(G) = \mathrm{SL}_2(5) \times Z_c$ for $p = 11$ and $c = 1$, $p = 29$ and $c = 7$, and $p = 59$ and $c = 29$. To see that these groups give Camina pairs, the reader will need to apply Lemma 4.30. We note that these groups are also Gagola groups. At this time, these are all of the nonsolvable Camina pairs that have been published.

We now want to consider what restrictions regarding nonsolvable Camina pairs and Gagola groups are known. Even though Camina and Gagola did not have any nonsolvable examples, they both obtained some restriction. In [12], Camina quotes results of W. B. Stewart in a preprint entitled "Largely fixed-point-free groups"to get information on nonsolvable Camina pairs. In a private communication, van der Waall has pointed out that Proposition 9.2 in Stewart's preprint is incorrect, and that Camina relies on this proposition to prove his corollary in Sect. 1 of [12], and so his proof of that result is incomplete. Gagola also obtains restrictions on the nonsolvable Gagola groups that can occur in Theorem 5.6 of [20].

Camina's results have been improved in Theorem 1.3 of [5] where they prove the following. Since the proofs in [5] do not rely on the work in [12], we do not need to be concerned about the problem mentioned by van der Waall in the proof of this result. If G is a group, we write G^∞ for the *solvable residual* of G. That is, G^∞ is the unique subgroup that is minimal subject to being normal in G and whose quotient is solvable.

Theorem 4.36 (Arad, Mann, Muzychuk, Pech) *Let (G, K) be a Camina pair where G is not solvable. Then, K is a p-group for a prime p and one of the following holds:*

1. $(G/\mathbf{O}_p(G))^\infty \cong \mathrm{SL}(2, p^e)$ *where e is an integer so that* $p^e > 3$.
2. $(G/\mathbf{O}_p(G))^\infty \cong \mathrm{SL}(2,5)$ *where* $p = 3$.
3. $(G/\mathbf{O}_p(G))^\infty \cong \mathrm{SL}(2, 13$ *where* $p = 3$.
4. $(G/\mathbf{O}_p(G))^\infty \cong \mathrm{SL}(2,5)$ *where $p \geq 7$ and (S, K) is a Camina pair so that G/S is isomorphic to either A_5 or S_5 where S is the solvable radical of G. That is, S is the unique subgroup that is maximal subject to being normal in G and solvable.*

Notice that, the example in [5] satisfies Conclusion (1) with $p = 5$, and the examples from [24] meet conclusion (4) of Theorem 4.36 with $p = 11, 29, 59$. Also, the nonsolvable Frobenius groups will meet Conclusion (4) for many other primes. Thus, it is reasonable to ask the following:

Open Question: Do there exist Camina pairs that satisfy Conclusions (2) and (3) of Theorem 4.36?

Open Question: Do there exist Camina pairs that satisfy Conclusion (1) with $p \neq 5$?

Open Question: For which primes p are there Camina pairs that satisfy Conclusion (4)?

4.6 Homogeneous Induction

We consider a class of groups studied by van der Waall and Kuisch in [37]. Let G be a group and let H be a subgroup of G. They say that (G, H) satisfies (CI) if every nonprincipal irreducible character of H induces homogeneously to G. We have seen in Lemma 4.1 that when H is normal this condition is equivalent to being a Camina pair.

When G is a p-group, then we have the following result which is Lemma 3.1 of [37].

Lemma 4.37 *Let G be a p-group for some prime p. If (G, H) satisfies (CI), then H is a term in the upper central series of G.*

In the situation of Lemma 4.37, we now have that H is normal in G, so Lemma 4.1 applies and (G, H) is a Camina pair. Next, we will see that the situation when H is not normal, we still have a Camina pair. The following is Theorem 3.2 of [37].

Theorem 4.38 (Kuisch, van der Waall) *Suppose (G, H) satisfies (CI). If N is the normal closure of H in G, then (G, N) is a Camina pair.*

More information can be obtained about the structure of H when (G, H) satisfies (CI) and H is not normal. The following are Theorems 4.1 and 4.7 of [37].

Theorem 4.39 (Kuishch, van der Waall) *If (G, H) satisfies (CI) and H is not normal, then H is nilpotent. Furthermore, if N is the normal closure of H, then either G is a doubly transitive Frobenius group whose Frobenius kernel is N or N is a p-group.*

In [70], they study pairs (G, H) that satisfy (CI) even further, focusing on the case where H is not normal. Their results are fairly technical, so we do not repeat them here except for the following. Suppose (G, H) satisfies (CI) where H is not normal in G and the normal closure of H is a p-group for some odd prime p and write N for the core of H in G. Then G/N has a Gagola character. They obtain a similar result with $p = 2$ under the additional hypothesis that G does not have a normal Sylow 2-subgroup (see Corollary 6.2 of [70]).

A generalization of homogeneous induction has been studied by Gagola and Sezer in [21] and Lyons in his dissertation [51]. They consider a group G that has a subgroup H with the property that every nonprincipal character of H induces to G as a sum of irreducible characters that all have the same degree. Obviously, if (G, H) satisfies CI, then G and H satisfy this condition. The results they obtain do not involve Camina pairs, so we will not digress further.

5 Generalizations

We now want to look at some of the generalizations of Camina groups and Camina pairs that have been studied.

5.1 Anti-central Elements

At this time, we look back to Lemma 2.1 that gave a number of equivalent conditions for an element of $g \in G$. In [38], Ladisch states that an element $g \in G$ is anti-central if it satisfies the conditions of Lemma 2.1. That is, g is anti-central if the conjugacy class of g is gG'. Note that these are the elements that we called Camina elements in the introduction. To avoid confusion with the rest of this paper, we are going to continue to refer to these elements as Camina elements as opposed to anti-central elements. However, we remind the reader that when they see anti-central element in [38] or other papers in the literature the definition is the same as our definition of Camina element.

Note that, G is a Camina group if and only if every element in $G \setminus G'$ is a Camina element. In [38], Ladisch considers any group that has a Camina element, and in that paper, he provides a number of examples of groups that have Camina elements that are not Camina groups. The main theorem of [38] is Theorem 4.3 which is the following.

Theorem 5.1 (Ladisch) *Let G be a (finite) group containing a Camina element. Then, G is solvable.*

In [38], Ladisch also gathers some interesting information regarding the structure of groups that contain a Camina element. For example, the following is Corollary 3.5 of [38].

Lemma 5.2 *Let N be a normal subgroup of G and let π be a set of primes. If $a \in G$ is a Camina element, then a fixes a unique Hall π-subgroup of N.*

The following is Proposition 3.6 of [38].

Lemma 5.3 *If G has a Camina element a and N is a normal subgroup of G contained in G' that has a cyclic Sylow p-subgroup, then N has a normal p-complement.*

For $g \in G$, following Ladisch, we define $C^i(g)$ inductively by $C^0(g) = 1$ and $C^{i+1}(g) = \{x \in G \mid [g, x] \in C^i(g)\}$. Let $C^\infty = \cup_{i \geq 0} C^i(g)$. The following is Theorem 3.9 of [38].

Theorem 5.4 (Ladisch) *Let $a \in G$ be a Camina element, and set $D = C^\infty(a)$. Then, the following are true:*

1. *D is a nilpotent self-normalizing subgroup of G.*
2. *$G = DG'$.*
3. *If $H \leq G$ so that $G = G'H$ and $a \in H$, then $D \leq H$.*
4. *If $K \leq G$ so that K is nilpotent and $a \in K$, then $K \leq D$.*
5. *If $L \leq G$ so that $G = G'L$, $a \in L$, and L is nilpotent, then $L = D$.*

Recall that a nilpotent subgroup of G that is self-normalizing is called a *Carter subgroup* of G. When G is a solvable group, it is known that G has a Carter subgroup and that all Carter subgroups are conjugate (Satz III.3.10 of [32]). Thus, Theorem 5.4 proves that if $a \in G$ is a Camina element, then $C^\infty(a)$ is a Carter subgroup of G.

Infinite groups having a Camina element have been studied in [19].

5.2 The Vanishing-Off Subgroup

Recall when G is a group and χ is a character of G that the *vanishing-off subgroup* of χ is the subgroup $V(\chi) = \langle g \in g \mid \chi(g) \neq 0 \rangle$. Thus, $V(\chi)$ is generated by the elements of G where χ does not vanish. That is, $V(\chi)$ is the smallest subgroup of G where χ vanishes on the elements outside of $V(\chi)$. Note that χ may also vanish on elements in $V(\chi)$.

When G is a nonabelian group, we defined in [41] the *vanishing-off subgroup* of G to be the subgroup of G generated by the elements $g \in G$ so that there exists a nonlinear irreducible character χ so that $\chi(g) \neq 0$. It is not hard to see that $V(G)$ is going to be the product of all the $V(\chi)$'s as χ runs over the nonlinear irreducible characters of G. In particular, every nonlinear irreducible character of G will vanish on all of the elements of G outside of $V(G)$ and $V(G)$ is the smallest subgroup of G with this property. Recall by Lemma 2.1 that an element $g \in G$ vanishes on every nonlinear irreducible character of G if and only if g is a Camina element. Thus, every element of $G - V(G)$ is a Camina element, and $V(G)$ is the smallest subgroup of G with the property that every element outside of it is a Camina element. Note that $V(G)$ may contain Camina elements, so $V(G)$ should be viewed as the subgroup generated by the non-Camina elements of G.

Notice that Theorem 5.1 implies that if G is nonsolvable, then $G = V(G)$. Also, it is not difficult to see that if G has nonlinear irreducible characters with coprime degrees, then $G = V(G)$ (see the first paragraph on p. 1314 of [41]). Thus, it is often the case that $G = V(G)$. However, there are interesting cases where $V(G) < G$. For example, G is a Camina group if and only if $G' = V(G)$. In [41], we studied the groups G where $V(G) < G$. The following is Theorem 1 of [41].

Theorem 5.5 *Let G be a nonabelian solvable group. Then, $G/V(G)$ is either cyclic or an elementary abelian p-group for some prime p.*

We then inductively define a central series for G in terms of $V(G)$. We set $V_1(G) = V(G)$ and for $i \geq 2$, we set $V_i(G) = [V_{i-1}(G), G]$. We use the following notation for the lower central series of G. We set $G_1 = G$ and for $i \geq 2$, we set $G_i = [G_{i-1}, G]$. Note that, this is consistent with our definition of G_3 earlier. The following is Theorem 2 of [41].

Theorem 5.6 *Let G be a group. Then, $G_{i+1} \leq V_i(G) \leq G_i$ for all $i \geq 1$. If in addition $V_2(G) < G_2$, then the following holds:*

1. *There is a prime p so that $G_{/}V_i(G)$ is an elementary abelian p-group for all $i \geq 1$.*
2. *There exist positive integers $m \leq n$ so that $|G : V(G)| = p^{2n}$, $|G_2 : V_2(G)| = p^m$, and $\text{cd}(G/V_2(G)) = \{1, p^n\}$.*

One should note the similarity to the result that if G is a semi-extra special p-group, then there exist integers $m \leq n$ so that $|G : G'| = p^{2n}$, $|G'| = p^m$, and $\text{cd}(G) = \{1, p^n\}$. When $V_3(G) < G_3$, Theorem 3 of [41] gave the next result.

Theorem 5.7 *Suppose G is a group where $V_3(G) < G_3$. Let $Z/V_3(G) = Z(G/V_3(G))$ and $C/V_3 = C_{G/V_3(G)}(G'/V_3(G))$. Then, the following are true:*

1. $|G : V_1(G)| = |G' : V_2(G)|^2$.
2. $V_2(G) = Z \cap G'$.
3. *Either* $|G : C| = |G' : V_2(G)|$ *or* $C = V_1(G)$.
4. $V(C) \leq V_1(G)$.
5. *If* $V_1(G) < C$, *then* $C' = V_2(G)$.
6. *If* $V_1(G) < C$ *and* $[V_1(G), C] < V_2(G)$, *then* $|G : C|$ *is a square.*

Again, note the similarities with the results regarding Camina groups of nilpotence class 3. We have seen that there are no Camina groups with nilpotence class more than 3. On the other hand, this central series associated with the vanishing-off subgroup seems to go further. In his dissertation [62] and in the preprint [63], our student Nabil Mlaiki studied groups G where $V_k(G) < G_k$ for an arbitrary positive integer k. He defined $D_k/V_k(G) = C_{G/V_k(G)}(G_{k-1}/V_k(G))$ for all integers $k \geq 3$. We now present Theorem 2 of [62, 63].

Theorem 5.8 (Mlaiki) *Suppose G is a group where $V_k(G) < G_k$ for some integer $k \geq 3$ and $G'/V_k(G)$ is abelian. Then, the following are true:*

1. $D_k = D_3$.
2. $|G_{k-1} : V_{k-1}(G)| = |G : D_3|$ *if* $k \geq 4$.
3. $|G_k : V_k(G)| \leq |G : D_3|$.

Note that $G_2 = G'$, so the subgroup D_3 in Theorem 5.8 equals the subgroup C in Theorem 5.7. We would like to understand what happens when one removes the hypothesis that $G'/V_k(G)$ is abelian. At this time, the only examples we have where $V_k(G) < G_k$ are examples found using the small groups library in the computer algebra system Magma. In particular, we do not have examples with $k \geq 5$. It would be nice to find families of groups that have examples with k arbitrarily large.

We close this section by mentioning an application of this group. Define the group G to be a *Q_1-group* if every nonlinear irreducible character has rational values. An element $g \in G$ is *rational in G* if $\chi(g)$ is rational for all $\chi \in \mathrm{Irr}(G)$. It is known that this is equivalent to g being conjugate in G to g^m for every integer m that is coprime to $o(g)$.

In Theorem 3.9 of [18], it is proved that G is a Q_1-group if and only if every element $g \in V(G)$ is rational in G. In [67], $V(G)$ is used in classifying the nonabelian Q_1-groups.

5.3 Camina Triples

In some sense, the most obvious way to generalize Camina pairs is in terms of Camina triples. Define (G, N, M) to be a *Camina triple* if $1 < M \leq N$ are proper normal subgroups of G so that for every element $g \in G \setminus N$, then g is conjugate to all of

gM. Notice that, (G, N, N) is a Camina triple if and only if (G, N) is a Camina pair. It is not difficult to see that $(G, V(G), G')$ will be a Camina triple for any group G where $V(G) < G$. In fact, if (G, N) is a Camina pair and M is a group, then $(G \times M, N \times M, N \times 1)$ will be a Camina triple. (See Theorem 1.2 of [30].)

This definition was originally considered by Mattarei in his dissertation (see Theorem 4.4.1 of [58]) where he was studying groups with the same character tables. We independently proposed this definition in [39]. Camina triples appeared in the paper [35] by Johnson, Mattarei, and Sehgal where they arise in the problem of weak Cayley tables. We will discuss the problems of groups with the same character tables and the same weak Cayley tables in a later section. Camina triples also arise in the question of fusions of character tables in [29, 30].

We now generalize the definition of the vanishing-off subgroup. Let N be a normal subgroup of G. We define $V(G \mid N)$ to be the product of the $V(\chi)$ where χ runs through the characters in $\mathrm{Irr}(G \mid N)$. In particular, it is easy to see that $V(G) = V(G \mid G')$. We first consider a couple of basic facts regarding this subgroup. This appears as Lemma 5.1 of [62].

Lemma 5.9 *If N is a normal subgroup of a group G, then $N \leq V(G \mid N)$.*

The following is Lemma 5.3 of [62].

Lemma 5.10 *Let G be a group. If M is a normal subgroup of G and $M \not\leq G'$, then $G = V(G \mid M)$.*

We now provide a connection between Camina triples and $V(G \mid N)$. We also obtain a generalization of Lemma 4.1 for Camina triples. The most of the following appears as Theorem 5.4 of [62] and as Theorem 2.1 of [64]. Condition (4) is from Theorem 1.3 of [30].

Theorem 5.11 *If $1 < M \leq N$ are two normal subgroups of G, then the following are equivalent:*

1. *(G, N, M) is a Camina triple.*
2. *$|C_G(g)| = |C_{G/M}(Mg)|$ for every $g \in G \setminus N$.*
3. *χ vanishes on $G \setminus N$ for every character $\chi \in \mathrm{Irr}(G \mid M)$.*
4. *Every character in $\mathrm{Irr}(N \mid M)$ induces homogeneously to G.*
5. *$V(G \mid M) \leq N$.*
6. *For all $g \in G \setminus N$ and $z \in M$, there exists an element $y \in G$ so that $[g, y] = z$.*

Notice that, an immediate corollary of this theorem is that (G, N) is a Camina pair if and only if $N = V(G \mid N)$. We also get the following observation regarding Camina triples. (See Lemma 2.2 of [64].)

Corollary 5.12 *If (G, N_1, M) and (G, N_2, M) are Camina triples, then $(G, N_1 \cap N_2, M)$ is a Camina triple. If (G, N, M) is a Camina triple and $N \leq K$, then (G, K, M) is a Camina triple.*

In Theorem 3 of [62] or Theorem 1 of [64], the following is proved. Note that, one consequence of Theorem 4.36 is that if (G, N) is a Camina pair where $N < G$, then N is solvable. Conclusion (a) shows that a generalization can be proved for Camina triples. Conclusion (b) is a generalization of Theorem 4.13 to Camina triples.

Theorem 5.13 (Mlaiki) *If (G, M, N) is a Camina triple such that $M < G$, then the following are true:*

1. *N is solvable.*
2. *If π is the set of primes dividing $|G : M|$, then N has a normal π-complement Q and N/Q is nilpotent.*

The following is Lemma 2.11 of [64].

Theorem 5.14 (Mlaiki) *If (G, M, N) is a Camina triple and G is nilpotent, then there is a prime p so that G/M and N are p-groups.*

In [30], a number of results are proved about Camina triples. The following is Theorem 1.2 of [30]. Notice that, this reduces the problem of finding Camina triples to groups that cannot be written as direct products.

Theorem 5.15 (Humphries, Kerby, Johnson) *Let G and H be groups.*

1. *If (G, J, K) is a Camina triple, then $(G \times H, J \times H, K \times 1)$ is a Camina triple.*
2. *Let $\pi : G \times H \to G$ be the projection map. Suppose that $(G \times H, J, K)$ is a Camina triple. If $\pi(K) \neq 1$, then $(G, \pi(J), \pi(K))$ is a Camina triple.*

5.4 More Generalizations

In [74], they define a group to be *flat* if every conjugacy class is a coset of a (necessarily normal) subgroup, and they consider p-groups with this property. (We note that this condition was considered for infinite groups in [8, 73].) It is not difficult to see that every nilpotent group of class 2 is flat. The following is Theorem 5.1 of [74].

Theorem 5.16 (Moran, Tandra) *Let G be a nilpotent group of class* 3.

1. *If G is a Camina group, then G is flat.*
2. *If $(G, Z(G))$ is a Camina pair, then G is flat.*
3. *If $|G| = p^5$, then $(G, Z(G))$ is a Camina pair if and only if G is flat.*

We see that Camina groups of nilpotence class 3 are flat. The paper [74] presents other p-groups of nilpotence class 3 that is flat. Using the work of Heineken in [26], one can show that all metacyclic groups are flat. Since the dihedral groups of 2-power order are metacyclic, we can find flat groups with arbitrarily large nilpotence class. (One of the referees has mentioned that there are also examples of flat 2-groups in [25] which is in Russian. At this time, we have not been able to obtain a copy of this paper ourselves, so we have not been able to verify this comment.)

In [22], they define a *waist* to be a normal subgroup W of a group G with the property that if N is any normal subgroup of G, then either $N \leq W$ or $W \leq N$. They note that if (G, N) is a Camina pair, then N must be a waist. They are mostly interested in waists in pro p-groups, and many of their results can be considered generalizations of results about Camina pairs. Some of their results are of particular interest in finite groups.

Theorem 5.17 (Gavioli, Monti, Scoppola) *Suppose p is an odd prime and G is a noncyclic p-group. Suppose W is a waist for G, then the following are true:*

1. $W \leq \Phi(G)$.
2. *G/W is not cyclic.*
3. *$|W| \neq p$, then W is a term in both the upper and lower central series for G.*

In his preprint [56], Mann notes that if W is a waist in G, then W must be the unique normal subgroup of G having order $|W|$. He also notes that in p-groups, the converse will also be true, but this need not be true otherwise.

In [65], the authors consider the following situations: (1) Suppose the group G has a normal subgroup N such that for every element $x \in G \setminus N$, the conjugacy class of x in G contains all of the elements of order $o(x)$ in $G \setminus N$ and (2) Suppose the group G has a normal subgroup M such that for every element $x \in G \setminus M$, the coset xM contains all of the elements of $o(x)$ in $G \setminus M$. In Proposition 4.3 of [65], it is proved that if G satisfies (1) and (2) with $M = N$, then $M = N = G'$ and G is a Camina group. Furthermore, in Proposition 4.4 of [65], they prove that if G is a Camina group, then G satisfies (1) with $N = G'$ if and only if G satisfies (2) with $M = G'$.

In the papers [6, 7], the authors define *auto Camina groups*. They define $G^* = \{g^{-1}g^{\alpha} \mid g \in G, \alpha \in \mathrm{Aut}(G)\} = [G, \mathrm{Aut}(G)]$. They say that G is an auto Camina group if $gG^* = \{g^{\alpha} \mid \alpha \in \mathrm{Aut}(G)\}$ for all $g \in G \setminus G^*$. In those papers, a few basic results regarding auto Camina groups are proved. In [6], they provide an example of an auto Camina group that is not a Camina group.

6 Applications

We are going to close this paper with a section outlining some of the problems where Camina groups and Camina pairs and their generalizations have been applied.

6.1 Identical Character Tables

We begin with character tables. Let G be a group. We label the irreducible characters of G as $\chi_1, \ldots, \chi_n$ and we can pick representatives $x_1, \ldots, x_n$ of the conjugacy classes of G. It is customary to fix $\chi_1 = 1_G$ and $x_1 = 1$. The *character table* of G is the $n \times n$ matrix whose i, j-entry is $\chi_i(x_j)$. In general, there is no canonical ordering of either the irreducible characters or the conjugacy classes. With this in mind, we say that two groups G_1 and G_2 have *identical* character tables if there exists some permutation of the rows and some permutation of the columns of the character table of G_2 so that its entries match up with the entries of the character table of G_1.

It is easy to see that isomorphic groups have identical character tables. On the other hand, it is not difficult to find nonisomorphic groups that have identical character tables. For example, if G_1 and G_2 are extra special groups of the same order, then G_1 and G_2 have identical character tables. In fact, if G_1 and G_2 are semi-extra special groups with $|G_1 : G_1'| = |G_2 : G_2'|$ and $|G_1'| = |G_2'|$, then G_1 and G_2 have identical character tables. This observation is made on p. 66 of [58] and later proved independently by us in [39]. In [40], we showed that if G_1 and G_2 are Camina groups of nilpotence class 3, then they have identical character tables if and only if $|G_1 : G_1'| = |G_2 : G_2'|$ and $|[G_1', G_1]| = |[G_2', G_2]|$. Observe that having identical character tables is an equivalence relation for groups.

One question that has received considerable attention is the question of what properties are determine by the character table of group. That is, if G_1 and G_2 are groups with identical character tables, what properties of G_1 does G_2 have to have. It is easy to see that if G_1 is abelian, then G_2 is abelian. It is somewhat more difficult to see that if G_1 is simple, then G_2 must be simple. If G_1 is nilpotent, then G_2 must be nilpotent. In fact, the upper central series can be read off of the character table, so if G_1 is nilpotent, then G_2 is nilpotent with the same nilpotence class.

It turns out that if G_1 and G_2 have identical character tables and G_1 is solvable, then G_2 must be solvable. It was quite a surprise when in his dissertation [58] and then in the papers [59–61], Mattarei presented examples of groups G and H that have identical character tables, but have different derived lengths. The groups G and H produced in [59] and Chap. 5 of [58] are semi-direct products of a q-group acting on a p-group. In both groups, the p-groups being acted on are Camina pairs. We note that in this case, the groups are not nilpotent and one is metabelian and the other has derived length 3.

In [60] and Chap. 6 of [58], the groups G and H are p-groups. Again one is metabelian and the other has derived length 3. In this case, both of the groups are themselves Camina pairs. In Corollary 4.3 of [58], there is a nice necessary condition for when two Camina pairs have identical character tables. In Theorem 4.4.1 of [58], this result is generalized to Camina triples. Finally, in [61] and Chap. 7 of [58], it is shown how to use wreath products to take these examples and obtain examples of groups G and H with identical character tables where one has derived length $d \geq 3$ and the other had derived length $d + 1$.

6.2 Weak Cayley Tables

The idea of a weak Cayley table was proposed in [35]. Let G be a group. We label the elements of G by $g_1 = 1, g_2, \ldots, g_n$. The *weak Cayley table* is the $n \times n$ matrix whose i, j-entry is the conjugacy class of $g_i g_j$. Weak Cayley tables were constructed to incorporate information from the character table of the group and the so-called "two-character" of the group. We are not going to discuss two characters here, but the interested reader can refer to the Introduction of [35] for an exposition on two-characters. Two groups G_1 and G_2 have *identical weak Cayley tables* if there is a bijection $\alpha : G_1 \to G_2$ such that (1) $g, h \in G_1$ are conjugate in G_1 if and only if $\alpha(g)$ and $\alpha(h)$ are conjugate in G_2 and (2) for all $g, h \in G_1$, $\alpha(gh)$ and $\alpha(g)\alpha(h)$ are conjugate in G_2. It is proved in Corollary 2.8 of [35] that if G_1 and G_2 have identical weak Cayley tables, then G_1 and G_2 have identical character tables. The following is Theorem 3.1 of [35].

Theorem 6.1 (Johnson, Mattarei, Sehgal) *Suppose (G_1, N) and (G_2, N) are Camina pairs so that N is an abelian group and $G_1/N \cong G_2/N$ has odd order. Then, G_1 and G_2 have identical weak Cayley tables.*

Notice that this implies that the groups constructed by Mattarei in [60] have identical weak Cayley tables, and so, the derived length of a group cannot be determined from its weak Cayley table. In [35], they explore a number of other connections between weak Cayley tables and Camina pairs.

6.3 Snyder's Problem

Let G be a nonabelian group and let d be the largest degree of a nonlinear irreducible character of G. Of course, we know that d divides $|G|$ and $|G| > d^2$. Thus, there exists a positive integer e so that $|G = d(d + e)$. Berkovich showed that $e = 1$ if and only if G is a 2-transitive Frobenius group (see [10]). Snyder noted when $e \geq 2$ that $|G|$ can be bounded in terms of e [72]. In [31], it was shown that $|G| \leq e^4 - e^3$ when $e > 1$ and that equality can only occur when G has a nontrivial abelian normal subgroup. It was previously known that examples exist that meet this bound; so this is the best possible bound. For the full details of this problem, see [31, 44] and the references therein.

For the purposes of this paper, we restrict ourselves to the case where G has a nontrivial abelian normal subgroup. This was the case considered in [44]. In that paper, the following is the key result.

Theorem 6.2 *Let G be a Gagola group and let N be the unique minimal normal subgroup. Write p for the prime so that N is an elementary abelian p-group. Assume G is not a 2-transitive Frobenius group with Frobenius kernel N, and let d and e be defined as above for G and note that d is the degree of the Gagola character. Then, $d \leq e^2 - e$. Furthermore, if P is a Sylow p-subgroup of G, then $|N|^2 \leq |P : N|$.*

Using Theorem 6.2, we then show that if G is any group with a nontrivial abelian normal subgroup with $e > 1$, then $d \leq e^2 - e$ where d and e is abelian. In Sect. 7 of [31], we consider groups G where $|G| = e^4 - e^3$, i.e., where $d = e^2 - e$. In Theorem 7.2 of [31], we prove the following.

Theorem 6.3 *Let G be a finite group, and let $|G| = d(d + e)$ where $d > 1$ is a character degree of G and $e > 1$ is an integer. Then, $|G| = e^4 - e^3$ if and only if G has a Gagola character of degree d and a unique minimal normal subgroup N of order e.*

On p. 545 of [31], we conjectured that $|G| = e^4 - e^3$ implies that G is solvable. However, if one considers the nonsolvable group G of order $2^3 \cdot 3 \cdot 5^6 = 375000$ with a normal subgroup N so that $|N| = 5^2 = 25$ that was found in [5] and mentioned in Sect. 4.5, then one sees that $e = 5^2 = 25$ and $d = 25 \cdot 24 = 600$. Hence, there exists a nonsolvable example where the bound is met, and thus, the conjecture in [31] is false.

6.4 Other Applications

In [28], Humphries and Johnson introduce the idea of fusion of character tables. The definition of fusing character tables is quite complicated, and so, we refer the interested reader to either [28] or [29] for the definition. In Theorem 1.1 of [28], they show that extra special p-groups fuse from an abelian group. In [29], they study which Camina pairs and Camina triples fuse from abelian groups. In particular, give an example of a group that fuses from an abelian group but contains a subgroup that does not fuse from any abelian group. The group with this property has a Camina triple. In [30], they prove that if G is a noncyclic group, then G fuses from a cyclic group if and only if G has normal subgroups $K \leq H$ so that (G, H, K) is a Camina triple that satisfies some additional conditions on H and K.

The problem where we first encountered Camina groups was the following. Let G be a solvable group and suppose that G has the property that if $a, b \in \mathrm{cd}(G)$ with $1 \neq a, b$ and $a \neq b$, then a does not divide b and b does not divide a. In other words, there is no divisibility among the nontrivial character degrees of G. This is the problem studied in [48]. In that paper, we prove that if G satisfies the nondivisibility hypothesis, then $|\mathrm{cd}(G)| \leq 4$ and the derived length of G is at most 3. Furthermore, we are able to describe the groups G in this set which has $|\mathrm{cd}(G)| = 4$. In particular, we show that if G is in this set and $|\mathrm{cd}(G)| = 4$ that G has a normal subgroup that is a Camina p-group that has nilpotence class 3. Note that, one of the key steps (Theorem 4.9 of [48]) reduces to showing that there exists no Camina group with nilpotence class greater than 3.

Camina groups and Camina pairs arise in the classification of groups with only one irreducible character of degree divisible by p. This classification is proved in [24]. In that paper, it is proved for a prime p that a group G has exactly one irreducible character whose degree is divisible by p if and only if G is one of the groups mentioned in a list of nine families of groups. We are not going to mention all of the groups here. Included in the list are extra special 2-groups and doubly transitive Frobenius groups whose Frobenius complements have a nontrivial cyclic normal Sylow p-subgroup. There are also two families that contain a normal Sylow p-

subgroup that is semi-extra special, and in one of these families there is a quotient that is a doubly transitive Frobenius group and in the other family, there is a subgroup that is a doubly transitive Frobenius group and we can apply Lemma 4.30 to see that the groups in this family is Camina pairs.

Dade and Yadav consider the question of groups where the product of any two conjugacy classes that do not contain inverses is another conjugacy class. Note that, this is a stronger condition than condition (5) of Lemma 4.1. They prove the following as Theorem A of [17]

Theorem 6.4 (Dade, Yadav) *Let G be a group. Then, the following are equivalent:*

1. *When C and D are conjugacy classes of G satisfying $x^{-1} \notin D$ for $x \in C$, then CD is a conjugacy class of G.*
2. *One of the following occurs:*
   (a) *G is a p-Camina group for some prime p,*
   (b) *G is a Frobenius group whose Frobenius kernel is elementary abelian of order p^n and a Frobenius complement is cyclic of order $p^n - 1$ where p is a prime and n is a positive integer so that $p^n > 2$.*
   (c) *G is a Frobeniu group whose Frobenius kernel is elementary abelian of order 9 and a Frobenius complement is the quaternion group of order* 8.

Looking at the question of the probability of that the product of two elements of a group G equals a fixed element $g \in G$ is the subject of [66]. In that paper, they study this question when G is nilpotent Camina group.

Chillag and Herzog consider groups with "almost distinct degrees" in [15]. In particular, they consider a group G with the property that χ, ψ being distinct nonlinear irreducible characters of G with $\chi(1) = \psi(1)$ implies that $\psi = \overline{\chi}$. To do this, they consider extended Camina pairs. They say (G, G') is an extended Camina pair if $1 < G' < G$ and xG' is contained in the union of the conjugacy classes of x and x^{-1}. They then show that if G is a nonabelian, nonperfect group so that (G, G') is an extended Camina pair, then one of the following holds: (1) G is a p-group for some prime p, (2) G is a Frobenius group with Frobenius kernel G', or (3) G/G' is a 2-group and $C_G(u)$ is a 2-group for every $u \in G \setminus G'$. Furthermore, if G is not a Camina group, then (3) holds. They also show that if G is a nonabelian, nonperfect group where $\chi(1) = \theta(1)$ for distinct nonlinear irreducible characters χ, θ only when $\chi = \overline{\theta}$, then (G, G') is an extended Camina pair. In [15], the authors use the fact that these groups are extended Camina pairs to obtain a classification. In [4], the authors consider the following variation: An *extended Camina group* is a pair (G, H) where H is a proper, nontrivial, normal subgroup of G so that Hg is contained in the union of the conjugacy classes of g and g^{-1} for all $g \in G \setminus H$.

A similar generalization of Camina groups is found in [49]. A group G is called a *relative elementary abelian group* (abbreviated *REA group*) if there is a subgroup $N < G$ such that if $g_1, g_2 \in G \setminus N$ then g_1 is conjugate under $\mathrm{Aut}(G)$ to either g_2 or g_2^{-1}. In [49], they study the Cayley graphs of these groups.

In [71], Ren uses Camina ideas to consider two problems. We say that G is a V^3-*group* if every nonlinear irreducible character takes on exactly three values.

Furthermore, G is said to be a *restricted* V^3-group if G is a V^3-group, and every nonlinear character $\chi \in \mathrm{Irr}(G)$ has its kernel contained in G'. Ren then proves that every V^3-group is solvable. He uses Camina techniques to classify the restricted V^3-groups. The classification includes semi-extra special 2-groups and Frobenius groups that satisfy some extra conditions.

Ren in [71] also considers D-groups. A group G is called a D-group if for every normal subgroup N satisfying $1 < N \leq G'$ and for every nonprincipal character $\lambda \in \mathrm{Irr}(N)$, the induced character λ^G has the property that distinct irreducible constituents have distinct degrees. Note that, D-groups had previously been studied in [9]. In [71], Ren uses the ideas of Camina groups to simplify the proofs and weaken the hypothesis for the classification of D-groups that are not perfect.

In [50], Loukaki showed that if G is a group with an abelian minimal normal subgroup N with the property that the irreducible characters of G that do not have N in their kernels have distinct degrees. That is, when $\chi, \psi \in \mathrm{Irr}(G)$ with $N \not\leq \ker(\chi)$, $N \not\leq \ker(\psi)$, and $\chi \neq \psi$, then it must be that $\chi(1) \neq \psi(1)$. Loukaki proves in this circumstance that (G, N) must be a Camina pair.

Zhang and Shi use Camina pairs in [79] when studying metabelian groups where the nonlinear irreducible characters each vanish on at most three conjugacy classes. In particular, they obtain a classification of these groups. They ask about what can be said if the hypothesis of metabelian is removed.

If G is a group, $\mathrm{Aut}_c(G)$ is the set of automorphisms of G so that $\alpha(x)$ is conjugate to x for all $x \in G$. It is not difficult to see that $\mathrm{Aut}_c(G)$ is a normal subgroup of $\mathrm{Aut}(G)$ that contains the inner automorphisms. In [78], Yadav proved that G is a p-group of order p^n, then $|\mathrm{Aut}_c(G)| \leq p^{(n^2-4)/4}$ when n is even and $|\mathrm{Aut}_c(G)| \leq p^{(n^2-1)/4}$ when n is odd. He also answers the question of when equality happens, and this involves consider Camina p-groups.

In [1, 23], it proved that if G is a p-group and $(G, Z(G))$ is a Camina pair, the G has a noninner automorphism of order p when p is odd or order 2 or 4 when $p = 2$ that fixes the subgroup $\Phi(G)$ elementwise.

In [41], we defined (G, N) to be a *generalized Camina pair (GCP)* if N is a normal subgroup of G and every element of $G \setminus N$ is a Camina element. It is not difficult to see that (G, N) is a GCP if and only if $(G, N, V(G))$ is a Camina triple. When $N = Z(G)$, it follows that $(G, Z(G))$ is a GCP if and only if every element of $G \setminus Z(G)$ is a Camina element. In [40], we defined G to be a *generalized Camina group* if every element in $G \setminus Z(G)$ is a Camina element. Thus, G is a generalized Camina group if and only if $(G, Z(G))$ is a GCP. In [68, 69], the authors study the total character of generalized Camina groups where the total character is the character $\sum_{\chi \in \mathrm{Irr}(G)} \chi$.

We close by mentioning that the idea of Camina groups and Camina pairs have been adapted to the setting of table algebras. Table algebras can be thought of as a generalization of finite groups. We suggest that the interested reader consult [2–4, 11] for more details. We are not going to pursue this further since the definitions get complicated and the topic is departure from group theory.

References

1. A. Abdollahi, S.M. Ghoraishi, On noninner 2-automorphisms of finite 2-groups. Bull. Aust. Math. Soc. **90**, 227–231 (2014)
2. Z. Arad, H.I. Blau, On table algebras and applications to finite groups. J. Algebra **138**, 137–185 (1991)
3. Z. Arad, E. Fisman, On table algebras, c-algebras, and applications to finite group theory. Commun. Algebra **19**, 2955–3009 (1991)
4. Z. Arad, J. Erez, E. Fisman, H.I. Blau, Real table algebras and applications to finite groups of extended Camina-Frobenius type. J. Algebra **168**, 615–647 (1994)
5. Z. Arad, A. Mann, M. Muzychuk, C. Pech, On non-solvable Camina pairs. J. Algebra **322**, 2286–2296 (2009)
6. H. Arora, R. Karan, A note on autocamina groups. Note Mat. **35**, 39–50 (2015)
7. M. Badrkhani Asl, M.R.R. Moghaddam, M.J. Sadeghifard, Some properties of auto-coset groups. Southeast Asian Bull. Math. **39**, 173–179 (2015)
8. L.W. Baggett, E. Kaniuth, W. Moran, Primitive ideal spaces, characters, and Kirillov theory for discrete nilpotent groups. J. Func. Anal. **150**, 175–203 (1997)
9. Y. Berkovich, Finite groups in which the degrees of some induced characters are distinct. Publ. Math. Debr. **44**, 225–234 (1994)
10. Y. Berkovich, Groups with few characters of small degree. Isr. J. Math. **110**, 325–332 (1999)
11. H.I. Blau, G. Chen, Reality-based algebras, generalized Camina-Frobenius pairs, and the nonexistence of degree maps. Commun. Algebra **40**, 1547–1562 (2012)
12. A.R. Camina, Some conditions which almost characterize Frobenius groups. Isr. J. Math. **31**, 153–160 (1978)
13. D. Chillag, I.D. MacDonald, Generalized Frobenius groups. Isr. J. Math. **47**, 111–122 (1984)
14. D. Chillag, A. Mann, C.M. Scoppola, Generalized Frobenius groups II. Isr. J. Math. **62**, 269–282 (1988)
15. D. Chillag, M. Herzog, Finite groups with almost distinct character degrees. J. Algebra **319**, 716–729 (2008)
16. R. Dark, C.M. Scoppola, On Camina groups of prime power order. J. Algebra **181**, 787–802 (1996)
17. E.C. Dade, M.K. Yadav, Finite groups with many product conjugacy classes. Isr. J. Math. **154**, 29–49 (2006)
18. M.R. Darafsheh, A. Iranmanesh, S.A. Moosavi, Groups whose non-linear irreducible characters are rational valued. Arch. Math. (Basel) **94**, 411–418 (2010)
19. K. Ersoy, Infinite groups with an anticentral element. Commun. Algebra **40**, 4627–4638 (2012)
20. S.M. Gagola Jr., Characters vanishing on all but two conjugacy classes. Pac. J. Math. **109**, 363–385 (1983)
21. S.M. Gagola Jr., S. Sezer, Induced characters with equal degree constituents. J. Group Theory **18**, 299–312 (2015)
22. N. Gavioli, V. Monti, C.M. Scoppola, Pro-p groups with waists. J. Algebra **351**, 130–137 (2012)
23. S.M. Ghoraishi, A note on automorphisms of finite p-groups. Bull. Aust. Math. Soc. **87**, 24–26 (2013)
24. D. Goldstein, R.M. Guralnick, M.L. Lewis, A. Moretó, G. Navarro, P.H. Tiep, Groups with exactly one irreducible character of degree divisible by p. Algebra Number Theory **8**, 397–428 (2014)
25. E.A. Golikova, A.I. Starostyn, Finite groups with uniquely generated normal subgroups, in "The subgroup structure of groups", Akad. Nauk. SSSR Ural, Otdel (1988)
26. H. Heineken, Commutator closed groups. Ill. J. Math. **9**, 242–255 (1965)
27. M. Herzog, P. Longobardi, M. Maj, On infinite Camina groups. Commun. Algebra **39**, 4403–4419 (2011)
28. S.P. Humphries, K.W. Johnson, Fusions of character tables and Schur rings of abelian groups. Commun. Algebra **36**, 1437–1460 (2008)

29. S.P. Humphries, K.W. Johnson, Fusions of character tables II: p-groups. Commun. Algebra **37**, 4296–4315 (2009)
30. S.P. Humphries, B.L. Kerby, K.W. Johnson, Fusions of character tables III: fusions of cyclic groups and a generalisation of a condition of Camina. Isr. J. Math. **178**, 325–348 (2010)
31. N.N. Hung, M.L. Lewis, A.A. Schaeffer Fry, Finite groups with an irreducible character of large degree. Manuscr. Math. **149**, 523–546 (2016)
32. B. Huppert, *Endliche Gruppen* (Springer, Berlin, 1983)
33. I.M. Isaacs, Coprime group actions fixing all nonlinear irreducible characters. Can. J. Math. **41**, 68–82 (1989)
34. I.M. Isaacs, M.L. Lewis, Camina p-groups that are generalized Frobenius complements. Arch. Math. (Basel) **104**, 401–405 (2015)
35. K.W. Johnson, S. Mattarei, S.K. Sehgal, Weak Cayley tables. J. Lond. Math. Soc. **61**(2), 395–411 (2000)
36. E.B. Kuisch, Sylow p-subgroups of solvable Camina pairs. J. Algebra **156**, 395–406 (1993)
37. E.B. Kuisch, R.W. van der Waall, Homogeneous character induction. J. Algebra **149**, 454–471 (1992)
38. F. Ladisch, Groups with anticentral elements. Commun. Algebra **36**, 2883–2894 (2008)
39. M.L. Lewis, in *Character tables of groups where all nonlinear irreducible characters vanish off the center*, Ischia Group Theory 2008 (World Scientific Publishing, Hackensack, 2009), pp. 174–182
40. M.L. Lewis, Generalizing Camina groups and their character tables. J. Group Theory **12**, 209–218 (2009)
41. M.L. Lewis, The vanishing-off subgroup. J. Algebra **321**, 1313–1325 (2009)
42. M.L. Lewis, On p-group Camina pairs. J. Group Theory **15**, 469–483 (2012)
43. M.L. Lewis, Classifying Camina groups: a theorem of Dark and Scoppola. Rocky Mountain J. Math. **44**, 591–597 (2014). See also: Erratum on Classifying Camina groups: a theorem of Dark and Scoppola [MR3240515]. Rocky Mountain J. Math. **45**, 273 (2015)
44. M.L. Lewis, Bounding group orders by large character degrees: a question of Snyder. J. Group Theory **17**, 1081–1116 (2014)
45. M.L. Lewis, Camina pairs that are not p-closed. Isr. J. Math. **206**, 89–94 (2015)
46. M.L. Lewis, Centralizers of Camina groups of nilpotence class 3. J. Group Theory **21**, 319–335 (2018)
47. M.L. Lewis, Semi-extraspecial groups, submitted
48. M.L. Lewis, A. Moretó, T.R. Wolf, Non-divisibility among character degrees. J. Group Theory **8**, 561–588 (2005)
49. C.H. Li, L. Wang, Relative elementary abelian groups and a class of edge-transitive Cayley groups. J. Aust. Math. Soc. **100**, 241–251 (2016)
50. M. Loukaki, On distinct character degrees. Isr. J. Math. **159**, 93–107 (2007)
51. C.F. Lyons, Induced characters with equal degree constituents, Ph.D. dissertation, Kent State University, 2016
52. I.D. Macdonald, Some p-groups of Frobenius and extra-special type. Isr. J. Math. **40**, 350–364 (1981)
53. I.D. Macdonald, More on p-groups of Frobenius type. Isr. J. Math. **56**, 335–344 (1986)
54. F. Mahmudi, A. Gholami, Some results on n-Camina pairs of groups. Ital. J. Pure Appl. Math. **36**, 887–898 (2016)
55. A. Mann, Some finite groups with large conjugacy classes. Isr. J. Math. **71**, 55–63 (1990)
56. A. Mann, More on normally monomial p-groups, Preprint
57. A. Mann, C.M. Scoppola, On p-groups of Frobenius type. Arch. der Math. **56**, 320–332 (1991)
58. S. Mattarei, Retrieving information about a group from its character table, Ph.D. Dissertation, University of Warwick, 1992
59. S. Mattarei, Character tables and metabelian groups. J. Lond. Math. Soc. (2) **46**, 92–100 (1992)
60. S. Mattarei, An example of p-groups with identical character tables and different derived lengths. Arch. Math. (Basel) **62**, 12–20 (1994)
61. S. Mattarei, On character tables of wreath products. J. Algebra **175**, 157–178 (1995)

62. N. Mlaiki, A central series associated with $V(G)$, Ph.D. dissertation, Kent State University, 2011
63. N. Mlaiki, A central series associated with $V(G)$, arXiv:1209.2886
64. N. Mlaiki, Camina triples. Can. Math. Bull. **57**, 125–131 (2014)
65. A.S. Muktibodh, S.H. Ghate, On Camina group and its generalizations. Mat. Vesnik **65**, 250–260 (2013)
66. R.K. Nath, M.K. Yadav, On the probability distribution associated to commutator word map in finite groups. Int. J. Algebra Comput. **25**, 1107–1124 (2015)
67. M. Norooz-Abadian, H. Sharifi, Sylow 2-subgroups of solvable Q_1-groups. C. R. Math. Acad. Sci. Paris **355**, 20–23 (2017)
68. S.K. Prajapati, B. Sury, On the total character of finite groups. Int. J. Group Theory **3**, 47–67 (2014)
69. S.K. Prajapati, R. Sarma, Total character of a group G with $(G, Z(G))$ as a generalized Camina pair. Can. Math. Bull. **59**, 392–402 (2016)
70. Y. Ren, X. Li, Z. Lu, On homogeneous character induction. Algebra Colloq. **3**, 355–368 (1996)
71. Y. Ren, Applications of the theory of Camina groups. Chin. Ann. Math. **20B**, 39–50 (1999)
72. N. Snyder, Groups with a character of large degree. Proc. Am. Math. Soc. **136**, 1893–1903 (2008)
73. H. Tandra, Characters of nilpotent groups, Ph.D. dissertation, Flinders University of South Australia, Adelaide, 2001
74. H. Tandra, W. Moran, Flatness conditions on finite p-groups. Commun. Algebra **32**, 2215–2224 (2004)
75. R.W. van der Waall, E.B. Kuisch, Homogeneous character induction II. J. Algebra **170**, 584–595 (1994)
76. L. Verardi, Gruppi semiextraseciali di esponente p. Ann. Mat. Pura Appl. **148**, 131–171 (1987)
77. H. Wang, X. Chang, P. Jin, On generalized Gagola characters. Arch. Math. (Basel) **104**, 501–508 (2015)
78. M.K. Yadav, Class preserving automorphisms of finite p-groups. J. Lond. Math. Soc. **75**, 755–772 (2007)
79. J.S. Zhang, W.J. Shi, A note on zeros of characters of finite groups. J. Math. Res. Expo. **28**, 589–592 (2008)

The Upper Central Series of the Unit Groups of Integral Group Rings: A Survey

Sugandha Maheshwary and Inder Bir S. Passi

MSC2000 16U60 · 16S34 · 20C05 · 20C07 · 20F14

1 Introduction

For a group G, let $\mathcal{U}(\mathbb{Z}G)$ be the group of units of the integral group ring $\mathbb{Z}G$ and let $\mathcal{V} := \mathcal{V}(\mathbb{Z}G)$ be the group of normalized units in $\mathbb{Z}G$, so that $\mathcal{U}(\mathbb{Z}G) = \pm\mathcal{V}$. The aim of this article is to survey results on the upper central series $\langle 1\rangle = \mathcal{Z}_0(\mathcal{V}) \subseteq \mathcal{Z}_1(\mathcal{V}) \subseteq \cdots \subseteq \mathcal{Z}_n(\mathcal{V}) \subseteq \mathcal{Z}_{n+1}(\mathcal{V}) \subseteq \ldots$ of $\mathcal{V}$.

In case G is finite, the central height of $\mathcal{V}$, i.e., the smallest integer $n \geq 0$ such that $\mathcal{Z}_n(\mathcal{V}) = \mathcal{Z}_{n+1}(\mathcal{V})$, is at most 2 [6, 7]. Furthermore, the central height of $\mathcal{V}$ is 2 if, and only if, G is a Q^* group, i.e., G has an element a of order 4 and an abelian subgroup H of index 2, which is not an elementary abelian 2-group, such that $G = \langle H, a\rangle$, $h^a := a^{-1}ha = h^{-1}$, for all $h \in H$ and $a^2 = b^2$, for some $b \in H$. Moreover, in this case, $\mathcal{Z}_2(\mathcal{V}) = T\mathcal{Z}_1(\mathcal{V})$, where $T = \langle b\rangle \oplus E_2$, E_2 being an elementary abelian 2-group. In all other cases, the central height must be 0 or 1. Thus, for finite groups, the problem of understanding the upper central series of $\mathcal{V}$ boils down to the study of $\mathcal{Z}(\mathcal{V}) := \mathcal{Z}_1(\mathcal{V})$, the centre of $\mathcal{V}$, which has been a topic of intensive research (see [47, 48, 60, 65, 73]).

We begin by reviewing, in Sect. 2, the results on the free rank of $\mathcal{Z}(\mathcal{V})$. The results related to the groups where $\mathcal{Z}(\mathcal{V})$ is trivial in the sense that $\mathcal{Z}(\mathcal{V}) = \mathcal{Z}(G)$,

S. Maheshwary · I. B. S. Passi
Indian Institute of Science Education and Research,
Mohali, Sector 81, Mohali 140306, Punjab, India
e-mail: sugandha@iisermohali.ac.in

I. B. S. Passi (✉)
Centre for Advanced Study in Mathematics, Panjab University, Chandigarh 160014, India
e-mail: ibspassi@yahoo.co.in

N. S. N. Sastry and M. K. Yadav (eds.), *Group Theory and Computation*,
Indian Statistical Institute Series, https://doi.org/10.1007/978-981-13-2047-7_9

are discussed in Sect. 3. The structure of non-trivial $\mathcal{Z}(\mathcal{V})$ is taken up in Sect. 4. In Sect. 5, we discuss the results on the upper central series of integral group rings of arbitrary (not necessarily finite) groups. We conclude this survey with a discussion on the hypercentral units of $\mathbb{Z}G$ in Sect. 6.

2 Free Rank of $\mathcal{Z}(\mathcal{U}(\mathbb{Z}G))$

Let G be a finite group. It is known ([67], Corollary 7.3.3) that

$$\mathcal{Z}(\mathcal{U}(\mathbb{Z}G)) = \langle -1 \rangle \times \mathcal{Z}(G) \times A_G = \pm\mathcal{Z}(\mathcal{V}), \tag{1}$$

where A_G is a free abelian group of finite rank, say $\rho(G)$. To determine the structure of $\mathcal{Z}(\mathcal{V})$, the computation of $\rho(G)$ is thus an essential requirement.

For an abelian group $G, \mathcal{U}(\mathbb{Z}G)$ was first investigated by Higman in [39] where he explored various properties of group rings and thus paved the way for some of the most important questions in the theory of group rings. A rank formula for $\rho(G)$, in case G is abelian, was given by Ayoub and Ayoub [8].

Theorem 1 (Abelian groups) *([8], Theorem 4) If G is a finite abelian group, then*

$$\rho(G) = \frac{1}{2}(|G| + n_2 - 2c + 1), \tag{2}$$

where $|G|$ denotes the order of the group G, n_2 is the number of elements of order 2 in G and c is the number of cyclic subgroups of G.

The above formula (2) for $\rho(G)$ was generalized to all finite groups, in terms of the number C_G of conjugacy classes in G, the number q_G of $\mathbb{Q}$-conjugacy classes of G (two elements $x,\ y \in G$ are said to be $\mathbb{Q}$-conjugate, if the cyclic subgroups $\langle x\rangle,\ \langle y\rangle$ are conjugate in G) and the number r_G of real classes in G (a conjugacy class C is said to be real, if $x^{-1} \in C$ for every $x \in C$). Ritter and Sehgal [70] proved that for a finite group G,

$$\rho(G) = \frac{1}{2}(C_G - 2q_G + r_G). \tag{3}$$

Let $r(G)$ and $q(G)$ denote, respectively, the number of simple components in the Wedderburn decomposition of the real group algebra $\mathbb{R}G$ and the rational group algebra $\mathbb{Q}G$. Observing that $r(G)$ equals the number of $\mathbb{R}$-conjugacy classes in G (two elements $x,\ y \in G$ are said to be $\mathbb{R}$-conjugate, if x is conjugate to y or y^{-1}) and $q(G)$ equals q_G ([22], Theorem 42.8), Ferraz [26] independently obtained an alternate expression for $\rho(G)$ as follows:

$$\rho(G) = r(G) - q(G). \tag{4}$$

As an application, the free rank $\rho(\mathcal{A}_n)$ for the alternating groups $\mathcal{A}_n$, in terms of partitions of n, was computed. It may be mentioned that $\rho(\mathcal{A}_n)$ had also been given, using the theory of Young tableaux, by Giambruno and Jespers [32].

The classification of the alternating groups $\mathcal{A}_n$ with $\rho(\mathcal{A}_n)$ at most 1, has been obtained as follows:

Theorem 2 (Alternating groups) *[5, 26]*

(i) $\rho(\mathcal{A}_n) = 0$ *if and only if* $n \in \{1, 2, 3, 4, 7, 8, 9, 12\}$, *and*
(ii) $\rho(\mathcal{A}_n) = 1$ *if and only if* $n \in \{5, 6, 10, 11, 13, 16, 17, 21, 25\}$.

Using the results of [26], Ferraz and Simón [30] computed the rank $\rho(G)$ for a finite metacyclic group G by computing the number of $\mathbb{R}$-conjugacy classes and $\mathbb{Q}$-conjugacy classes of G. In particular, the following result was obtained:

Theorem 3 (Metacyclic groups) *Let $C_{p,q}$ be the non-abelian metacyclic group of order pq, where p and q are odd primes such that p divides $q-1$. Then,*

$$\rho(C_{p,q}) = \frac{p-1}{2} + \frac{q-1}{2p} - 2. \tag{5}$$

It is interesting to note that, the only solution of $\rho(C_{p,q}) = 0$ is $p = 3,\ q = 7$.

Recently, Jespers et al. [53] gave a formula to compute $\rho(G)$ for a large class of groups, including abelian-by-supersolvable groups. In order to present the same, we first recall the definition of strongly monomial groups [64].

Let G be a finite group and K a normal subgroup of a subgroup H of G. Define $\hat{H} := \frac{1}{|H|}\sum_{h\in H} h$ and

$$\varepsilon(H, K) := \begin{cases} \hat{H}, & \text{if } H = K; \\ \prod(\hat{K} - \hat{L}), & \text{otherwise,} \end{cases}$$

where $|H|$ denotes the order of H and L runs over the normal subgroups of H, which are minimal among the normal subgroups of H containing K properly. A *strong Shoda pair* [64] of G is a pair (H, K) of subgroups of G satisfying:

(i) K is normal in H and H is normal in $\mathcal{N}_G(K)$, the normalizer of K in G;
(ii) H/K is cyclic and a maximal abelian subgroup of $\mathcal{N}_G(K)/K$; and
(iii) the distinct G-conjugates of $\varepsilon(H, K)$ are mutually orthogonal.

It is known that if (H, K) is a strong Shoda pair of G, then $e(G, H, K)$, the sum of distinct G-conjugates of $\varepsilon(H, K)$, is a primitive central idempotent of the rational group algebra $\mathbb{Q}G$ ([64], Proposition 3.3). Two strong Shoda pairs (H_1, K_1) and (H_2, K_2) are said to be *equivalent*, if $e(G, K_1, H_1) = e(G, K_2, H_2)$. By *a complete irredundant set* of strong Shoda pairs of G, one means a complete set of representatives of the distinct equivalence classes of strong Shoda pairs of G. A finite group G is said to be *strongly monomial*, if every primitive central idempotent of $\mathbb{Q}G$ is of the form $e(G, H, K)$ for some strong Shoda pair (H, K) of G.

Jespers et al [53], calculated the free rank of $\mathcal{Z}(\mathcal{U}(\mathbb{Z}G))$ in terms of strong Shoda pairs of G, provided G is a strongly monomial group.

Theorem 4 (Strongly monomial groups) *Let G be a finite strongly monomial group. Then,*

$$\rho(G) = \sum_{(H,K)} \left(\frac{\varphi([H : K])}{k_{(H,K)}[\mathcal{N}_G(K) : H]} - 1 \right), \tag{6}$$

where $[A : B]$ denotes the index of B in A, φ is the Euler totient function, the sum runs over a complete irredundant set of strong Shoda pairs (H, K) of G with $H/K = \langle hK \rangle$ and

$$k_{(H,K)} = \begin{cases} 1, & \text{if } hh^x \in K \text{ for some } x \in \mathcal{N}_G(K); \\ 2, & \text{otherwise.} \end{cases}$$

For illustrations on the computation of rank using the above theorem, see ([12], Sect. 4), where $\rho(G)$ has been computed for non-abelian groups of orders p^3 and p^4, p-prime.

3 CUT-Groups

Trivially, $\mathcal{Z}(\mathcal{V})$ contains $\mathcal{Z}(G)$, the centre of G. In case $\mathcal{Z}(\mathcal{V}) = \mathcal{Z}(G)$, i.e., all central units are trivial, following [13], we call G a cut-*group*, or a group with the cut-*property*. Clearly, for a finite group G, $\mathcal{V}$ has central height zero if, and only if, G is a cut-group with trivial centre.

The question of classifying cut-groups was explicitly posed, for the first time, by Goodaire and Parmenter [33]. As an answer, Ritter and Sehgal [68] gave a characterization of such finite groups in terms of their conjugacy classes, which was later generalized for arbitrary groups [24]. Recently, cut-groups have further been explored [9, 13, 20, 63]. It turns out that the study of cut-groups has been going on under different names, and with different approaches. The various notions thus developed exhibit a highly interesting interplay between group theory, representation theory, algebraic number theory and K-theory.

In view of ([9], Proposition 2.2), ([14], Theorem 20.2), ([24], Lemma 2), ([43], p. 545), ([52], Corollary 1.7) and the results in [9, 20, 26, 68], we have the following characterizations for a finite group to be a cut-group:

Theorem 5 *The following statements are equivalent for a finite group G:*

(i) *G is a cut-group.*
(ii) *The free rank $\rho(G)$ of $\mathcal{Z}(\mathcal{V})$ equals 0.*
(iii) *For every x in G, and for every natural number j, relatively prime to $|G|$, the order of G, x^j is conjugate in G to x or x^{-1}.*

(iv) *G is an inverse semi-rational group, i.e., each element $x \in G$ is such that every generator of the cyclic group $\langle x \rangle$ is conjugate in G to x or x^{-1}.*
(v) *The numbers $r(G)$ and $q(G)$, denoting respectively the number of simple components in the Wedderburn decomposition of $\mathbb{R}G$ and $\mathbb{Q}G$, are equal.*
(vi) *The character field $\mathbb{Q}(\chi) := \mathbb{Q}(\{\chi(g) \mid g \in G\})$ of each absolutely irreducible character χ of G is either $\mathbb{Q}$ or an imaginary quadratic field.*
(vii) *If $\mathbb{Q}G \cong \bigoplus_i M_{n_i}(\mathbb{D}_i)$, is the Wedderburn decomposition of $\mathbb{Q}G$, where $M_n(\mathbb{D})$ denotes the algebra of $n \times n$ matrices over the division ring $\mathbb{D}$, then the centre $\mathcal{Z}(\mathbb{D}_i)$ of each division ring $\mathbb{D}_i$ is $\mathbb{Q}$ or an imaginary quadratic field.*
(viii) *$G = \mathcal{N}_{\mathcal{V}}(G)$, the normalizer of G in $\mathcal{V}$.*
(ix) *$K_1(\mathbb{Z}G)$, the Whitehead group of $\mathbb{Z}G$, is finite.*

The above characterizations of cut-groups immediately yield the following properties of such groups (see [13, 63, 68]).

Proposition 1 *Let G be a finite* cut-*group. Then,*

(i) *Every homomorphic image $\overline{G}$ of G is a* cut-*group.*
(ii) *The centre $\mathcal{Z}(G)$ of G is a* cut-*group.*
(iii) *If H is a real group (i.e., for all $h \in H$, h is conjugate to h^{-1}), and has the* cut-*property, equivalently, if H is a rational group (i.e., for every $h \in H$, all generators of the cyclic subgroup $\langle h \rangle$ are in one conjugacy class of G), then the direct sum $G \oplus H$ is a* cut-*group.*

It is well known [39] that a finite abelian group G is a cut-group if, and only if, the exponent of G is $1, 2, 3, 4$ or 6. In fact, it is easy to check that if a finite group G has exponent $1, 2, 3, 4$ or 6, then it is a cut-group. However, the converse is not true for non-abelian groups. For instance, the non-abelian metacyclic group of order 27 is a cut-group and has an element of order 9 (see Theorem 7).

It may be noted that the cut-property is not direct sum closed. For instance, let

$$H = \langle a,\ b \mid a^8 = b^2 = 1,\ b^{-1}ab = a^3 \rangle \text{ and } K = \langle x \mid x^4 = 1 \rangle.$$

Then $H \oplus K$ is not a cut-group, although both H and K are cut-groups ([13], Remark 1). Further, it may be pointed that the statement (iii) in Proposition 1 is a generalization of Higman's result [39] which states that if G is a cut-group, then so is the direct sum $G \oplus E_2$, E_2 an elementary abelian 2-group.

In Theorem 4, for a strong Shoda pair $(H,\ K)$ of a strongly monomial finite group G with $[H : K] = m$ say, the quotient group $\mathcal{N}_G(K)/H$ is regarded as a subgroup of $\mathcal{U}(\mathbb{Z}/m\mathbb{Z})$, using the following faithful action:

$$\begin{aligned} \mathcal{N}_G(K)/H &\longrightarrow \mathrm{Gal}(\mathbb{Q}(\zeta_m)/\mathbb{Q})\ (\cong \mathcal{U}(\mathbb{Z}/m\mathbb{Z})) \\ xH &\longmapsto \alpha_{xH},\ x \in \mathcal{N}_G(K), \end{aligned}$$

where ζ_m is a primitive mth root of unity, $\alpha_{xH}(\zeta_m) = \zeta_m^j$, if $h^x K = h^j K$. With this identification, Theorem 4 at once yields the following:

Theorem 6 (Strongly monomial cut-groups) *[12] A strongly monomial group G is a* cut-*group if, and only if,*

$$\mathcal{U}(\mathbb{Z}/[H : K]\mathbb{Z}) = \langle \mathcal{N}_G(K)/H, -1\rangle,$$

for every strong Shoda pair (H, K) *of* G. *In particular, if for every strong Shoda pair* (H, K) *of a strongly monomial group* G,

$$[H : K] = 1, 2, 3, 4 \text{ or } 6,$$

then G *is a* cut*-group.*

An advantage of determining the primitive central idempotent $e(G, H, K)$ of $\mathbb{Q}G$, using strong Shoda pair (H, K) of G, is that one can describe the structure of the corresponding simple component $\mathbb{Q}Ge(G, H, K)$ ([64], Proposition 3.4).

The following theorem gives, up to isomorphism, a complete list of metacyclic cut-groups. This has been obtained by invoking Theorem 5 and by computing the structure of simple components of the rational group algebras of metacyclic groups via complete irredundant sets of strong Shoda pairs of G determined by the work in [11].

Theorem 7 (Metacyclic cut-groups) *([13], Theorem 5) Let* G *be a finite non-abelian metacyclic group defined by the presentation*

$$G = \langle a, b \mid a^n = 1, b^t = a^\ell, a^b = a^r\rangle,$$

where n, t, r, ℓ *are natural numbers such that*

$$r^t \equiv 1 \ (\text{mod } n), \ \ell r \equiv \ell \ (\text{mod } n) \text{ and } \ell \mid n.$$

Then, G *has the* cut*-property if, and only if,* G *is isomorphic to one of the following 46 groups:*

$\langle a, b \mid a^n = 1, b^t = 1, a^b = a^{n-1}\rangle$, $t = 2, 4, 6$, $n = 3, 4, 6$;
$\langle a, b \mid a^4 = 1, b^t = a^2, a^b = a^3\rangle$, $t = 2, 6$;
$\langle a, b \mid a^n = 1, b^{\varphi(n)} = 1, a^b = a^{\lambda_n}\rangle$, $n = 5, 7, 9, 10, 14, 18$;
$\langle a, b \mid a^n = 1, b^{\varphi(n)} = 1, a^b = a^{\lambda_n^2}\rangle$, $n = 7, 9, 14, 18$;
$\langle a, b \mid a^n = 1, b^{\frac{\varphi(n)}{2}} = 1, a^b = a^{\lambda_n^2}\rangle$, $n = 7, 9$;
$\langle a, b \mid a^8 = 1, b^t = 1, a^b = a^r\rangle$, $t = 2, 4$, $r = 3, 5$;
$\langle a, b \mid a^{12} = 1, b^t = 1, a^b = a^5\rangle$, $t = 2, 4$;
$\langle a, b \mid a^{12} = 1, b^6 = a^\ell, a^b = a^7\rangle$, $\ell = 6, 12$;
$\langle a, b \mid a^{15} = 1, b^4 = 1, a^b = a^2\rangle$;
$\langle a, b \mid a^{16} = 1, b^4 = 1, a^b = a^r\rangle$, $r = 3, 5$;
$\langle a, b \mid a^{20} = 1, b^4 = 1, a^b = a^r\rangle$, $r = 3, 13$;
$\langle a, b \mid a^{20} = 1, b^4 = a^{10}, a^b = a^3\rangle$;
$\langle a, b \mid a^{21} = 1, b^6 = 1, a^b = a^r\rangle$, $r = 2, 10$;
$\langle a, b \mid a^{28} = 1, b^6 = a^\ell, a^b = a^{11}\rangle$, $\ell = 14, 28$;
$\langle a, b \mid a^{30} = 1, b^4 = 1, a^b = a^{17}\rangle$;
$\langle a, b \mid a^{36} = 1, b^6 = a^\ell, a^b = a^7\rangle$, $\ell = 6, 36$;
$\langle a, b \mid a^{42} = 1, b^6 = 1, a^b = a^r\rangle$, $r = 11, 19$;

where λ_n *is a generator of* $\mathcal{U}(\mathbb{Z}/n\mathbb{Z})$.

For a classification of finite metacyclic groups G, relative to the central height of $\mathcal{V}(\mathbb{Z}G)$, see ([13], Theorem 6).

The following result extends Higman's classification of finite abelian cut-groups [39] to that of finite nilpotent cut-groups.

Theorem 8 (Nilpotent cut-groups) *([63], Theorem 3; see also [13], Sect. 2) A finite nilpotent group G has the cut-property if, and only if, G is one of the following:*

(i) *a 2-group such that for all $x \in G$, x^3 is conjugate to x or x^{-1};*
(ii) *a 3-group such that for all $x \in G$, x^2 is conjugate to x^{-1};*
(iii) *a direct sum $H \oplus K$ of a real group H satisfying* (i) *and a non-trivial group K satisfying* (ii).

Corollary 1 (p-groups) *Let G be a p-group with the cut-property. Then, $p = 2$ or 3 and each quotient in the lower, as well as the upper, central series of G is of exponent*

$$\begin{cases} 2 \text{ or } 4, & \text{if } p = 2; \\ 3, & \text{if } p = 3. \end{cases}$$

Corollary 2 *Let H and K be p-groups with the cut-property. Then, the following statements holds:*

(i) *$H \oplus K$ is a cut-group, if $p = 3$.*
(ii) *If $p = 2$, and one of H or K is a real group, then $H \oplus K$ is a cut-group.*

From the foregoing analysis and using the results from Sect. 4 of [12], it is easy to deduce the following identification of non-abelian cut-groups of order p^3 and p^4, p-prime.

Proposition 2 *A non-abelian group G of order p^3 or p^4, p-prime, is a cut-group if, and only if, it is isomorphic to one of the following 12 groups:*

$$\langle a, b : a^4 = b^2 = 1, ba = a^3 b\rangle,$$
$$\langle a, b : a^4 = 1,\ a^2 = b^2, ba = a^3 b\rangle,$$
$$\langle a, b : a^8 = b^2 = 1, ba = a^5 b\rangle,$$
$$\langle a, b, c : a^4 = b^2 = c^2 = 1,\ cb = a^2 bc,\ ab = ba,\ ac = ca\rangle,$$
$$\langle a, b : a^4 = b^4 = 1,\ ba = a^3 b\rangle,$$
$$\langle a, b, c : a^4 = b^2 = c^2 = 1,\ ca = a^3 c,\ ba = ab,\ cb = bc\rangle,$$
$$\langle a, b, c : a^4 = b^2 = c^2 = 1,\ ca = abc,\ ba = ab,\ cb = bc\rangle,$$
$$\langle a, b, c : a^4 = b^4 = c^2 = 1,\ ba = a^3 b,\ ca = ac,\ cb = bc,\ a^2 = b^2\rangle,$$
$$\langle a, b : a^8 = b^2 = 1,\ ba = a^3 b\rangle,$$
$$\langle a, b, c : a^9 = b^3 = c^3 = 1,\ ca = a^4 c,\ ba = ab,\ cb = bc\rangle,$$
$$\langle a, b, c : a^9 = b^3 = c^3 = 1,\ ba = a^4 b,\ ca = abc,\ cb = bc\rangle,$$
$$\langle a, b, c, d : a^3 = b^3 = c^3 = d^3 = 1,\ dc = acd,\ bd = db,\ ad = da,$$
$$bc = cb,\ ac = ca,\ ab = ba\rangle.$$

Specializing to p-groups of class 2, we have the following:

Proposition 3 (p-groups of class 2) *[63] Let G be a p-group of class 2. Then, the following statements are equivalent:*

(i) *G is a* cut*-group.*
(ii) *Either* (a) $p = 2$ *and for every* $x \in G$, $x^4 \in [x, G]$ *or*
(b) $p = 3$ *and, for every* $x \in G$, $x^3 \in [x, G]$, *where* $[x, G] := \langle x^{-1}x^g \mid g \in G\rangle$.
(iii) *For all $x \in G$, both $[x, G]$ and $G/[x, G]$ are* cut*-groups.*
(iv) *For all normal subgroups N of G contained in $\mathcal{Z}(G)$, both N and G/N are* cut*-groups.*

Furthermore, the cut*-property is direct sum closed for p-groups of class 2.*

Recently, Bächle [9] has given the classification of Frobenius cut-groups.

Theorem 9 (Frobenius cut-groups) *([9], Theorem 1.3) Let K be a Frobenius complement.*

(i) *If $|K|$ is even and K is the complement of a Frobenius* cut*-group G, then G is isomorphic to one of the groups in either $(a) - (f)$ with $b, c, d \in \mathbb{Z}_{\geq 1}$ or $(\alpha) - (\delta)$:*

(a) $C_3^b \rtimes C_2$; (α) $C_5^2 \rtimes Q_8$;

(b) $C_3^{2b} \rtimes C_4$; (β) $C_5^2 \rtimes (C_3 \rtimes C_4)$;

(c) $C_3^{2b} \rtimes Q_8$; (γ) $C_5^2 \rtimes SL(2, 3)$;

(d) $C_5^c \rtimes C_4$; (δ) $C_7^2 \rtimes SL(2, 3)$.

(e) $C_7^d \rtimes C_6$;

(f) $C_7^{2d} \rtimes (Q_8 \times C_3)$;

Conversely, for each of the above structure descriptions, there is a Frobenius cut*-group of that form, and it is unique up to isomorphism.*
(Here, C_n^m denotes the direct sum of m copies of the cyclic group C_n.)

(ii) *If $|K|$ is odd, then there is a Frobenius* cut*-group G with complement K and Frobenius kernel F if, and only if, $K \cong C_3$ and one of the following holds:*

(a) *F is a* cut*-2-group admitting a fixed point free automorphism of order 3. In particular, F has order 2^a for some $a \in 2\mathbb{Z}_{\geq 1}$ and is an extension of an abelian group of exponent a divisor of 4 by an abelian group of exponent a divisor of 4.*
(b) *F is an extension of an elementary abelian 7-group by an elementary abelian 7-group, exponent of F equals 7 and F admits a fixed point free automorphism of order 3 fixing each cyclic subgroup of F.*

Corollary 3 ([9], Corollary 4.6) *Let G be a Frobenius* cut*-group with abelian Frobenius kernel. Then, G appears in Theorem 9(i) or is isomorphic to*

$$(C_2^a \times C_4^{a'}) \rtimes C_3 \text{ or } C_7^d \rtimes C_3 \quad (a, a' \in 2\mathbb{Z}_{\geq 0},\ a + a' > 0,\ d \in \mathbb{Z}_{\geq 1}).$$

For each of the above structures, there is a unique Frobenius cut*-group of that form.*

Recall that a group G is called a *Camina group*, if $G \neq G'$, the derived subgroup of G, and, for every $g \notin G'$, the coset gG' is a conjugacy class [23, 62].

Bakshi et al. [13] have proved that non-abelian Camina p-groups are cut-groups, for $p = 2,\ 3$. As a corollary to classification of Frobenius cut-groups, the following is a classification of Camina cut-groups:

Corollary 4 (Camina cut-groups) *([9], Corollary 4.8) A Camina group G is a* cut-*group if, and only if,*

(i) G is a p-group, $p = 2,\ 3$,
(ii) G is a Frobenius group of the form:

$$(C_2^{2n} \times C_4^{2m}) \rtimes C_3,\ C_3^n \rtimes C_2,\ C_3^{2n} \rtimes C_4,\ C_3^{2n} \rtimes Q_8,\ C_5^2 \rtimes Q_8,$$

for $m, n \in \mathbb{Z}_{\geq 1}$,
(iii) G is a Frobenius group of the form:

$$C_7^n \rtimes C_3,\ C_5^n \rtimes C_4,\ C_7^n \rtimes C_6,$$

for $n \in \mathbb{Z}_{\geq 1}$, where a generator of the complement raises each element of the Frobenius kernel to the same power, or
(iv) G is a Frobenius cut*-group with a cyclic complement of order 3 and non-abelian kernel as described in Theorem 9 (ii).*

The following results from [63] use Theorem 8 along with the results on semi-rational groups due to Chillag and Dolfi [20].

Theorem 10 (Solvable cut-groups) *([63], Theorem 2) A finite solvable group G in which every element has prime-power order is a* cut*-group if, and only if, every element $x \in G$ satisfies one of the following conditions:*

(i) *$o(x) = 2^a$, $a \geq 0$ and x^3 is conjugate to x or x^{-1};*
(ii) *$o(x) = 7$ or 3^b, $b \geq 1$ and x^5 is conjugate to x^{-1};*
(iii) *$o(x) = 5$ and x^3 is conjugate to x^{-1};*

where $o(x)$ denotes the order of x.

The odd-order cut-groups have been definitively characterized as follows:

Theorem 11 (Odd-order cut-groups) *([63], Theorem 1) An odd-order group G is a* cut*-group if, and only if, every element $x \in G$ satisfies*

(i) *x^5 is conjugate to x^{-1}, and*
(ii) *$o(x)$ is either 7, or a power of 3.*

dummy

An immediate consequence of the above result is the following:

Corollary 5 *The* cut*-property is direct sum closed for odd-order groups.*

Crucial for the proof of the Theorem 11 is the observation that if G is an odd-order cut-group, then every element of G has prime-power order.

The characterization of even-order solvable but non-nilpotent cut*-groups having an element of mixed order is still an open problem.*

We conclude this section with the observation that if G is solvable, then the cut-property has a strong bearing on the prime spectrum $\pi(G)$, the set of primes dividing the order of G.

Theorem 12 (Prime spectrum of cut-groups) *[9, 13, 20, 63] Let G be a finite* cut*-group. Then, either 2 or 3 $\in \pi(G)$, and*

(i) *$\pi(G) \subseteq \{2, 3\}$, if G is nilpotent;*
(ii) *$\pi(G) \subseteq \{3, 7\}$, if G is an odd-order group;*
(iii) *$\pi(G) \subseteq \{2, 3, 5, 7\}$, if G is solvable.*

The bounds on $\pi(G)$ given in the above result are best possible, in the sense that no prime can be dropped. Since all symmetric groups S_n, $n \geq 3$ are cut-groups [68], such a restriction cannot be put on $\pi(G)$ if G is non-solvable.

The investigation of the cut*-property for non-solvable groups will naturally be of interest.*

4 The Structure of $\mathcal{Z}(\mathcal{U}(\mathbb{Z}G))$

If G is a finite group which is not a cut-group, then, for a complete description of $\mathcal{Z}(\mathcal{U}(\mathbb{Z}G))$, one needs to compute a basis S_G say, of a complement A_G of its torsion subgroup $\mathcal{Z}(G)$ (see (1)). Apparently, this involves construction of units of infinite order. While the construction of such units is rather hard, certain standard procedures have been evolved for cyclic group rings, which we proceed to recall (see [73] for details).

Let $C_n = \langle g \rangle$, be a cyclic group of order $n \geq 1$.

Bass cyclic units: Let k and m be positive integers with $1 < k < n$ and $k^m \equiv 1 (\mathrm{mod}\ n)$, then the element

$$u_{k,m}(g) := (1 + g + \cdots + g^{k-1})^m + \frac{1 - k^m}{n}(1 + g + \cdots + g^{n-1}) \tag{7}$$

is a unit in $\mathbb{Z}C_n$ with the inverse $u_{k',m}(g^k)$, where $kk' \equiv 1(\mathrm{mod}\ n)$, $1 < k' < n$.

Alternating units: Let c be an integer coprime to $2n$, then

$$\mu := 1 - g + g^2 - \cdots + g^{c-1}, \tag{8}$$

is a unit in $\mathbb{Z}C_n$.
Hoeschmann units: For $i \geq 0$ and $y \in C_n$, set

$$s_i(y) = 1 + y + \cdots + y^{i-1}. \tag{9}$$

Let i, j be integers, $0 < i, j < n$, both relatively prime to n and let k, l be positive integers with $li = 1 + kn$. Then,

$$u_{i,j}(g) = s_l(g^i)s_i(g^j) - ks_n(g) \tag{10}$$

is a unit in $\mathbb{Z}C_n$.

Set of Generators for $\mathcal{Z}(\mathcal{U}(\mathbb{Z}G))$

The problem of determining an explicit basis S_G of a complement of $\mathcal{Z}(G)$ in $\mathcal{Z}(\mathcal{U}(\mathbb{Z}G))$ has been solved for very few cases. We proceed to present the current status.

Cyclic Groups

If $n = 1, 2, 3, 4$ or 6, then C_n is a cut-group and therefore, $S_{C_n} = \phi$.
For C_5, we have $S_{C_5} = \{u\}$, where

$$u = g - g^3 - g^4,$$

and for C_8, we have $S_{C_8} = \{v\}$, where

$$v = 2 + g - g^3 - g^4 - g^5 + g^7 \quad ([67, 73]).$$

Aleev and Panina [3] proved that

$$\mathcal{V}(\mathbb{Z}G) = G \times \langle u_1 \rangle \times \langle u_2 \rangle, \text{ if } G = C_7 \text{ or } C_9,$$

and provided explicit description of the elements u_i, $i = 1, 2$.

The group $\mathcal{V}(\mathbb{Z}C_n)$, for $n = 10$ and 12 has been described in [1]. Aleev and Sokolev [4] gave the description of $\mathcal{V}(\mathbb{Z}G)$ when $G = C_{16}$ or C_{32}. Further, for a prime p satisfying certain number-theoretic condition, a basis S_{C_p}, was computed in [27], including, in particular, all primes $5 \leq p \leq 67$. The description of the elements of S_{C_p} was given in terms of *Bass cyclic units* and *alternating units*. Using similar conditions and methodology, Ferraz and Katani [28] extended the work in [27] to the case when $G = C_{p^m}$, involving *Hoeschmann units*. Their method is for restricted values of m and includes all primes p and $m \in \mathbb{N}$ satisfying $\varphi(p^m) \leq 66$. Recently, Ferraz and Marcuz [29] described a basis S_G when G is the cyclic group of order $2p$.

Metacyclic Groups

Adapting the algorithm given in ([18], Sect. 2.5.3), Jespers and del Río ([48], Example 7.2.4), illustrate the calculation of a basis S_G, when

$$G = \langle a, b \mid a^5 = 1, b^4 = 1, a^b = a^{-1} \rangle,$$

involving some GAP [75] computations. It has been pointed out that the algorithm is effective, but not too efficient in general ([21], Conclusion 4.9.3). The same technique can be used to find an S_G, when

$$G = \langle a, b \mid a^{13} = 1, b^4 = 1, a^b = a^{-5} \rangle,$$

D_{16} (the dihedral group of order 16), Q_{16} (the generalized quaternion group of order 16), or $\mathcal{A}_5$ ([48], Exercises 7.2.4 and 7.2.5). In [29], Ferraz and Marcuz described an S_G, when $G = C_p \times C_2 \times C_2$, for a prime p satisfying some suitable conditions. Furthermore, for certain primes p and q, Ferraz and Simón [31] described the structure of $\mathcal{Z}(\mathcal{V}(\mathbb{Z}C_{p,q}))$, with $C_{p,q}$ as in Theorem 3.

Alternating Groups

The group of central units of integral group ring $\mathbb{Z}\mathcal{A}_n$ of the alternating group $\mathcal{A}_n$ is known only for some values of n. The cases for which $\rho(\mathcal{A}_n) = 0$ (i.e., $S_G = \phi$) have already been given in Theorem 2. The generators of non-trivial groups $\mathcal{Z}(\mathcal{V}(\mathbb{Z}\mathcal{A}_n)), n = 5$ and $n = 6$, were explicitly given by Aleev [1]. For $\mathcal{A}_5$, same work was also independently carried out by Li and Parmenter [56]. The full description of $\mathcal{Z}(\mathcal{V}(\mathbb{Z}\mathcal{A}_n))$, for the cases when $\rho(\mathcal{A}_n) = 1$ (see Theorem 2), can be found in [2, 4]. Some results on calculation of $\rho(\mathcal{A}_n), n \leq 600$, can be found in [5].

Large Subgroups of $\mathcal{Z}(\mathcal{U}(\mathbb{Z}G))$

As may be noted from the preceding discussion, the problem of determining a multiplicatively independent subset yielding a torsion-free complement of $\mathcal{Z}(G)$ in $\mathcal{Z}(\mathcal{U}(\mathbb{Z}G))$ is not answered fully even for cyclic groups of prime order. A weaker, but still non-trivial, question is to find a *large subgroup of* $\mathcal{Z}(\mathcal{U}(\mathbb{Z}G))$, meaning a subgroup of finite index in $\mathcal{Z}(\mathcal{U}(\mathbb{Z}G))$. A lot of work has been done in this direction which we now proceed to describe.

For a cyclic group G, Bass cyclic units generate a subgroup of finite index in $\mathcal{U}(\mathbb{Z}G)$ [15]. Generalizing this result, Bass, Milnor, and Serre [16] proved that the result holds good for finite abelian groups as well. Their proof makes use of K-theory in order to reduce the computation to group rings of cyclic groups. In [54], Jespers et al gave another construction of subgroup of finite index in $\mathcal{U}(\mathbb{Z}G)$, for an abelian group G, which did not involve the use of K-theory. For abelian groups G, the units of $\mathbb{Z}G$ have been studied in a series of papers [40–42]. In ([41], Theorem 2.5), it is proved that the Hoechsmann units arising from the cyclic subgroups of order greater than 2, along with $\pm G$, generate a subgroup of finite index in $\mathcal{U}(\mathbb{Z}G)$.

Giambruno and Jespers [32] described a construction of a large subgroup in $\mathcal{Z}(\mathcal{U}(\mathbb{Z}\mathcal{A}_n))$. Their construction avoids the use of Bass cyclic units and is based mainly on the theory of Young tableaux. Ferraz and Simón [30], gave a subgroup of finite index in $\mathcal{Z}(\mathcal{U}(\mathbb{Z}C_{p,q}))$, by defining two kinds of units, both based on Bass cyclic units.

For an arbitrary finite group G, a construction of generators of a large subgroup in $\mathcal{Z}(\mathcal{U}(\mathbb{Z}G))$ can be found in [69]. In [51], Jespers et al provided an explicit set of generators for a large subgroup in $\mathcal{Z}(\mathcal{U}(\mathbb{Z}G))$, G a finitely generated nilpotent group. For a finite abelian-by-supersolvable group G, such that every cyclic subgroup of order not a divisor of 4 or 6 is subnormal in G, Jespers et al [50] gave a large subgroup in $\mathcal{Z}(\mathcal{U}(\mathbb{Z}G))$.

Based on Bass cyclic units and theory of strong Shoda pairs of G, Jespers and Parmenter [49] introduced a new construction of units which yield a subgroup of finite index in $\mathcal{Z}(\mathcal{U}(\mathbb{Z}G))$, provided G is strongly monomial group. With this construction, Jespers, Olteanu, del Río and Van Gelder [50] defined *generalized Bass units* as follows:

Let M be a normal subgroup of G and let $g \in G$ be an element of order n. Let k and m be positive integers with $1 < k < n$ and $k^m \equiv 1 \pmod{n}$, then

$$u_{k,m}(1 - \hat{M} + g\hat{M}) := 1 - \hat{M} + u_{k,m}(g)\hat{M} \tag{11}$$

is a unit in $\mathbb{Z}G(1-\hat{M}) + \mathbb{Z}G\hat{M}$. Since both $\mathbb{Z}G(1-\hat{M}) + \mathbb{Z}G\hat{M}$ and $\mathbb{Z}G$ are orders in $\mathbb{Q}G$, there is a positive integer $n_{g,M}$ such that

$$(u_{k,m}(1 - \hat{M} + g\hat{M}))^{n_{g,M}} \in \mathcal{U}(\mathbb{Z}G). \tag{12}$$

Suppose $n_{G,M}$ is the minimal positive integer satisfying (12) for all $g \in G$. Then, the element

$$(u_{k,m}(1 - \hat{M} + g\hat{M}))^{n_{G,M}} = 1 - \hat{M} + u_{k,mn_{G,M}}(g)\hat{M} \tag{13}$$

is called a *generalized Bass unit* of $\mathbb{Z}G$ based on g and M with parameters k and m.

Observe that $n_{G,M} = 1$, if M is trivial, i.e., $M = \langle 1 \rangle$ or G, and with $M = \langle 1 \rangle$, $u_{k,m}(1 - \hat{M} + g\hat{M}) = u_{k,m}(g)$, a Bass cyclic unit in $\mathbb{Z}G$.

Let G' denote the derived subgroup of G.

With the foregoing notation, we have

Theorem 13 ([50], Theorem 5.1) *Let G be a finite strongly monomial group. Then, the group generated by the generalized Bass units* $(u_{k,m}(1 - \hat{H}' + h\hat{H}'))^{n_{H,H'}} \in \mathbb{Z}H$, *arising from strong Shoda pairs* (H, K) *of G with* $H/K = \langle hK \rangle$, *contains a subgroup of finite index in* $\mathcal{Z}(\mathcal{U}(\mathbb{Z}G))$.

Note that, the preceding Theorem is a generalization of ([49], Corollary 2.3), where this result was proved for the class of metabelian groups. In ([49], Corollary 3.3; see also [60], Theorem 3.6), for a Frobenius group G with complement H of odd order, a large subgroup in $\mathcal{Z}(\mathcal{U}(\mathbb{Z}G))$ has been given.

Virtual Basis of $\mathcal{Z}(\mathcal{U}(\mathbb{Z}G))$

We next discuss the work, where not only a large subgroup in $\mathcal{Z}(\mathcal{U}(\mathbb{Z}G))$ for a finite group G has been constructed, but also a *virtual basis*, i.e., a multiplicatively independent set of elements of $\mathcal{Z}(\mathcal{U}(\mathbb{Z}G))$, generating such a subgroup has been given.

For a cyclic group, Bass described a virtual basis using the Independence Lemma [15]. However, for abelian groups, virtual basis was given much later in [54].

If G is a finite abelian-by-supersolvable group, such that every cyclic subgroup of order not a divisor of 4 or 6 is subnormal in G, a virtual basis has been provided in [50].

Recently, for a class of strongly monomial groups G which have a complete irredundant set of strong Shoda pairs (H, K) with the property that the index $[H : K]$ is a prime-power, Jespers et al [53] gave a virtual basis, say $\mathcal{B}$, of $\mathcal{Z}(\mathcal{U}(\mathbb{Z}G))$. The groups which satisfy this property include metacyclic groups $C_{q^m} \rtimes C_{p^n}$, where p and q are different odd primes, and C_{p^n} acts faithfully on C_{q^m}. Hence, this was an extension of work in [30], where a virtual basis was given for $\mathcal{Z}(\mathcal{U}(\mathbb{Z}C_{p,q}))$.

Index of Large Subgroups in $\mathcal{Z}(\mathcal{U}(\mathbb{Z}G))$

For abelian groups, Bass cyclic units generate a subgroup of finite index which is relatively higher in comparison to the index of the subgroup generated by Hoechsmann units. In [25], Faccin et al. provide an algorithm based on construction of Hoechsmann units, and implement the same in MAGMA [19], to compute the index for all groups of order up to 110. Furthermore, for the class of strongly monomial groups, studied by Jespers et al in [53], Bakshi and Maheshwary [12] estimated the index

$$[\mathcal{Z}(\mathcal{U}(\mathbb{Z}G)) : \langle \mathcal{B} \rangle]$$

of the free abelian subgroup generated by $\mathcal{B}$. An upper bound on the index has been given, based on the ideas contained in [53] and Kummer's work (see [76], Theorem 8.2) on the index of cyclotomic units.

Note that, index estimation for the large subgroups in full unit group of $\mathbb{Z}G$ has been done for certain other cases. For details, the reader is referred to Section 7 of [47].

We next consider central units in integral group rings of infinite groups.

5 Integral Group Rings of Infinite Groups

For an arbitrary abelian group G, it was shown by Sehgal ([71], Theorem 1, see also [72], Theorem 3.5), that every unit $u \in \mathbb{Z}G$ can be written as $u = wg,\ g \in G,\ w \in \mathbb{Z}T$, where T is a finite subgroup of G. This result was extended to central units of $\mathbb{Z}G$, when G is a finitely generated nilpotent group, by Jespers, Parmenter and Sehgal [51] and was later generalized further by Milies and Sehgal [66], to the case of arbitrary groups.

Recall that the FC-subgroup $\Phi(G)$ of a group G is the subgroup consisting of all elements in G having only finitely many conjugates in G. We denote by $\Phi^+(G)$, the torsion subgroup of $\Phi(G)$.

Theorem 14 ([66], Theorem 1) *Let G be an arbitrary group. Every central unit u of $\mathbb{Z}G$ can be written as*

$$u = wg = gw,\ g \in \Phi(G),\ w \in \mathbb{Z}\Phi^+(G). \tag{14}$$

The above theorem serves as a major tool to construct large subgroups of $\mathcal{Z}(\mathcal{U}(\mathbb{Z}G))$, using the generators of large subgroups in $\mathcal{Z}(\mathcal{U}(\mathbb{Z}\Phi^+(G)))$, when $\Phi(G)$ is finitely generated (see e.g. [51, 66]).

Dokuchaev, Milies and Sehgal [24] generalized, to arbitrary groups, the criterion given by Ritter and Sehgal (Theorem 5 (iv)), for a finite group to have the cut-property.

Theorem 15 ([24]) *An arbitrary group G is a* cut*-group if, and only if, every finite normal subgroup A of G satisfies the condition that, for every $a \in A$ and every natural number j relatively prime to the order of a, the element a^j is conjugate (in G) to a or a^{-1}.*

The representation for central units given in Theorem 14 was generalized to that for normalizing units $u \in \mathcal{N}_{\mathcal{V}}(G)$ and applied to verify the *normalizer property*, for several classes of infinite groups.

Theorem 16 ([52], Theorem 1) *Let G be an arbitrary group. Every normalizing unit $u \in \mathcal{N}_{\mathcal{V}}(G)$ can be written as*

$$u = wg,\ g \in G,\ w \in \mathbb{Z}\Phi^+(G). \tag{15}$$

As a consequence, it was shown that the groups $\mathcal{N}_{\mathcal{V}}(G)/G$ and $\mathcal{Z}(\mathcal{V})/\mathcal{Z}(G)$ are embedded in the torsion-free abelian group $\mathcal{Z}(\mathcal{V}) \cap \mathcal{V}(\mathbb{Z}\Phi^+(G))/\mathcal{Z}(G) \cap \Phi^+(G)$ and they all have the same torsion-free rank ([52], Corollary 1.5). One thus has another characterization of the cut-property for arbitrary groups.

Theorem 17 ([52], Corollary 1.7) *The following statements are equivalent for an arbitrary group G:*

(i) *G is a* cut*-group.*
(ii) $\mathcal{N}_{\mathcal{V}}(G) = G$.
(iii) *all units in $\mathbb{Z}\Phi^+(G)$, that are central in $\mathbb{Z}G$ are trivial.*

Finally, we consider the hypercentral units of integral group rings.

6 The Hypercentre of $\mathcal{V}(\mathbb{Z}G)$

For a group H, let

$$\mathcal{Z}_\infty(H) := \cup_{n=1}^{\infty} \mathcal{Z}_n(H)$$

denote the hypercentre of H. The group H is said to be *hypercentral* if $\mathcal{Z}_\infty(H) = H$. Given a group G, the elements of $\mathcal{Z}_\infty(\mathcal{V}(\mathbb{Z}G))$ are called *hypercentral units.*

Hypercentral Unit Groups

The classification of groups G for which $\mathcal{Z}_n(\mathcal{V}) = \mathcal{V}$ for some $n \geq 1$, i.e., $\mathcal{V}$ is nilpotent, due to Sehgal and Zassenhaus ([74], see also [72], VI.3.23), has long been known. It is naturally of interest to know when is $\mathcal{Z}_\infty(\mathcal{V}) = \mathcal{V}$, i.e., when is $\mathcal{V}$ hypercentral. Such groups were characterized by Bist [17] and have also been recently studied by Iwaki and Juriaans [44–46], using a different approach.

Theorem 18 ([17], Corollary 3, see also [46], Theorem 2.4) *Let G be an arbitrary group. The unit group $\mathcal{V}(\mathbb{Z}G)$ is hypercentral if and only if G is hypercentral and the torsion subgroup T of G satisfies one of the following conditions:*

(i) *T is central in G.*
(ii) *T is an abelian, non-central and for $g \in G$ and for every $t \in T$, $t^g = t$ or t^{-1}.*
(iii) *T is a Hamiltonian 2-group and every subgroup of T is normal in G.*

Hypercentral Units

While the central units of integral group rings have been a subject of intensive research, hypercentral units too have received considerable attention. However, unlike central units, no general constructions for hypercentral units seem to be known. The investigations on hypercentral units of integral group rings have been primarily motivated by the study of the *multiplicative Jordan decomposition* and the *normalizer property*.

A group G is said to have the *multiplicative Jordan decomposition property*, if every unit $u \in \mathcal{V}$ is expressible as the product $u = \alpha\beta$ with α semisimple, β unipotent and $\alpha\beta = \beta\alpha$ (see surveys [34, 35]). In the study of Jordan decomposition for elements of $\mathcal{V}$, Arora, Hales and Passi [6, 7] proved that, for a finite group G, $[\mathcal{Z}_2(\mathcal{V}), \mathcal{V}] \subseteq \mathcal{Z}(G)$, and concluded that $\mathcal{Z}_\infty(\mathcal{V}) = \mathcal{Z}_2(\mathcal{V}) = \mathcal{Z}(\mathcal{V})T$, where T is the torsion subgroup of $\mathcal{Z}_2(\mathcal{V})$. Moreover, for a finite group G, $\mathcal{Z}_2(\mathcal{V}) \neq \mathcal{Z}_1(\mathcal{V})$ if, and only if, G is a Q^* group.

In a series of papers, Li and Parmenter made significant contributions to the study of hypercentral units [55, 57–59]. They extended the above mentioned results of Arora, Hales and Passi to torsion groups and checked that $\mathcal{Z}_\infty(\mathcal{V}) = \mathcal{Z}_2(\mathcal{V})$ does not hold good in general [58, 59]. They also proved that the inclusion $\mathcal{Z}_\infty(\mathcal{V}) \subseteq G.\mathcal{Z}(\mathcal{V})$ holds if either (i) G is an FC-group which is locally nilpotent or has no 2-torsion; or (ii) the torsion elements of G form an abelian subgroup T contained in the FC-subgroup of G and $G = \langle T, g\rangle$ for some $g \in G$; or (iii) $G = T \rtimes X$, where T is finite abelian and X is torsion-free abelian ([58], Proposition 2.5). If the torsion elements of a group G form a subgroup T, say, then ([59], Theorem 3.2, see also [58], Theorem 2.3) implies that if $\mathcal{Z}_\infty(\mathcal{V}) \subseteq \mathcal{N}_\mathcal{V}(G)$, then $\mathcal{Z}_\infty(\mathcal{V}) \subseteq G.\mathcal{C}_\mathcal{V}(T)$, where $\mathcal{C}_\mathcal{V}(T)$ is the centralizer of T in $\mathcal{V}$. Furthermore, if $\mathcal{Z}_\infty(\mathcal{V}) \nsubseteq \mathcal{C}_\mathcal{V}(T)$, then T is either an abelian 2-group or T has an element a of order 4 and an abelian subgroup H of index 2, which is not an elementary abelian 2-group, such that $T = \langle H, a\rangle$, $a^{-1}ha = h^{-1}$, for all $h \in H$ ([59], Theorem 3.1, see also [58], Theorem 3.5).

A group G is said to have *the normalizer property*, if $\mathcal{N}_\mathcal{V}(G) = \mathcal{Z}(\mathcal{V}).G$. The normalizer property is related to some central problems in the theory of group rings and has been widely studied by several authors (see survey [10], for instance). The study of hypercentral units is closely related to the study of the normalizer property as well.

Theorem 19 ([38], Proposition 4.1, see also [59], Lemma 2 and [55], Lemma 1) *For an arbitrary group G,*

(i) $\mathcal{Z}_\infty(\mathcal{V}) \subseteq \mathcal{N}_\mathcal{V}(G)$;
(ii) $[\mathcal{V}, \mathcal{Z}_{n+1}(\mathcal{V})] \subseteq \mathcal{Z}_n(G)$, *for each* $n \in \mathbb{N}$;
(iii) *each element of* $\mathcal{Z}_\infty(\mathcal{V})$ *commutes with all the unipotent elements of* $\mathcal{V}$.

It may be noted that the above result has been proved more generally for the group ring RG, where R is a G-adapted ring (an integral domain R of characteristic zero is said to be G-adapted, if every rational prime p for which G has an element of order p, is not invertible in R). For partial or complete generalizations of results in this section, to RG, we refer the reader to [36].

Furthermore, it has been proved that the normal closure of group generated by the support of a hypercentral unit in $\mathcal{Z}_\infty(\mathcal{V}(\mathbb{Z}G))$ is a polycyclic-by-finite group, provided the group G is finitely generated ([38], Proposition 2.4).

In view of Theorem 19, the normalizer property clearly has a strong impact on the hypercentral units; for, then $\mathcal{Z}_\infty(\mathcal{V}) \subseteq G.\mathcal{Z}(\mathcal{V})$. Considerable work has been done in this direction [36–38, 45, 46, 52, 61].

Hertweck and Jespers [37] proved that Blackburn groups have the normalizer property. As a consequence of this, along with results in [38] (Propositions 4.1 and 4.5, Corollaries 4.3 and 4.12), it is observed that the inclusion $\mathcal{Z}_\infty(\mathcal{V}) \subseteq G.\mathcal{Z}(\mathcal{V})$ holds for any arbitrary group G.

Theorem 20 ([37]) *Let G be an arbitrary group. Then,*

$$\mathcal{Z}_\infty(\mathcal{V}) \subseteq G.\mathcal{Z}(\mathcal{V}).$$

Returning to the cut-groups, we see that the cut-property has strong bearing on the hypercentral units. The following result was proved by Li ([55], Corollary 2) for torsion groups.

Corollary 6 ([52], Propostion 1.8) *Let G be an arbitrary* cut*-group. Then,*

$$\mathcal{Z}_\infty(\mathcal{V}) \subseteq G.$$

Acknowledgements The authors are thankful to the anonymous referee(s) for their valuable comments and suggestions.
Support from Indo-Russian DST-RSF Project INT/RUS/RSF/P-2 is gratefully acknowledged.
The first author thankfully acknowledges the research support by DST, India (PDF/2016/000731; INSPIRE/04/2017/000897).
The second author would like to thank the organizers of the International Conference on *Group Theory and Computational Methods* held at International Centre for Theoretical Sciences, Tata Institute of Fundamental Research, Bangalore, during November 2016, for providing the opportunity to participate and enjoy their warm hospitality. Thanks are also due to Indian National Science Academy, New Delhi (India), for their support and Ashoka University, Sonipat (India), for making available their facilities.

References

1. R.Zh. Aleev, Higman's central unit theory, units of integral group rings of finite cyclic groups and Fibonacci numbers. Int. J. Algebra Comput. **4**(3), 309–358 (1994)
2. R.Zh. Aleev, The units of character fields and the central units of integer group rings of finite groups [translation of Mat. Tr. **3**(1, 3–37) (2000); MR1778756 (2001g:16062)], Siberian Adv. Math. **11**(1), 1–33
3. R.Zh. Aleev, G.A. Panina, The units of cyclic groups of orders 7 and 9, Izv. Vyssh. Uchebn. Zaved. Mat. no. 11, 81–84 (1999)
4. R.Zh. Aleev, V.V. Sokolov, On central unit groups of integral group rings of alternating groups. Proc. Steklov Inst. Math. **267**(suppl. 1), S1–S9 (2009)
5. R.Zh. Aleev, A.V. Kargapolov, V.V. Sokolov, The ranks of central unit groups of integral group rings of alternating groups. Fundam. Prikl. Mat. **14**(7), 15–21 (2008)
6. S.R. Arora, I.B.S. Passi, Central height of the unit group of an integral group ring. Commun. Algebra **21**(10), 3673–3683 (1993)
7. S.R. Arora, A.W. Hales, I.B.S. Passi, Jordan decomposition and hypercentral units in integral group rings. Commun. Algebra **21**(1), 25–35 (1993)
8. R.G. Ayoub, C. Ayoub, On the group ring of a finite abelian group. Bull. Austral. Math. Soc. **1**, 245–261 (1969)

9. A. Bächle, Integral group rings of solvable groups with trivial central units. Forum Math. **30**(4), 845–855 (2018)
10. A. Bächle, A survey on the normalizer problem for integral group rings.. Groups St Andrews 2013. Selected papers of the conference, St. Andrews, UK, August 3–11 (2013), Cambridge University Press, Cambridge (2015), pp. 152–159 (English)
11. G.K. Bakshi, S. Maheshwary, The rational group algebra of a normally monomial group. J. Pure Appl. Algebra **218**(9), 1583–1593 (2014)
12. G.K. Bakshi, S. Maheshwary, On the index of a free abelian subgroup in the group of central units of an integral group ring. J. Algebra **434**, 72–89 (2015)
13. G.K. Bakshi, S. Maheshwary, I.B.S. Passi, Integral group rings with all central units trivial. J. Pure Appl. Algebra **221**(8), 1955–1965 (2017)
14. H. Bass, K-theory and stable algebra. Inst. Hautes Études Sci. Publ. Math. no. 22, 5–60 (1964)
15. H. Bass, The Dirichlet unit theorem, induced characters, and Whitehead groups of finite groups. Topology **4**, 391–410 (1965)
16. H. Bass, J. Milnor, J.P. Serre, Solution of the congruence subgroup problem for SL_n $(n \geq 3)$ and Sp_{2n} $(n \geq 2)$. Inst. Hautes Études Sci. Publ. Math. no. 33, 59–137 (1967)
17. V. Bist, Unit groups of integral group rings. Proc. Am. Math. Soc. **120**(1), 13–17 (1994)
18. A.I. Borevich, I.R. Shafarevich, *Number theory*, Translated from the Russian by Newcomb Greenleaf. Pure and Applied Mathematics, vol. 20 (Academic Press, New York-London, 1966)
19. W. Bosma, J. Cannon, C. Playoust, The Magma algebra system. I. The user language. J. Symbolic Comput. **24**, no. 3-4, 235–265 (1997), Computational algebra and number theory (London, 1993)
20. D. Chillag, S. Dolfi, Semi-rational solvable groups. J. Group Theory **13**(4), 535–548 (2010)
21. H. Cohen, *A Course in Computational Algebraic Number Theory*, vol. 138, Graduate Texts in Mathematics (Springer, Berlin, 1993)
22. C.W. Curtis, I. Reiner, *Representation theory of finite groups and associative algebras* (AMS Chelsea Publishing, Providence, 2006). Reprint of the 1962 original
23. R. Dark, C.M. Scoppola, On Camina groups of prime power order. J. Algebra **181**(3), 787–802 (1996)
24. M. Dokuchaev, C. Polcino Milies, S.K. Sehgal, Integral group rings with trivial central units II. Commun. Algebra **33**(1), 37–42 (2005)
25. P. Faccin, W.A. de Graaf, W. Plesken, Computing generators of the unit group of an integral abelian group ring. J. Algebra **373**, 441–452 (2013)
26. R.A. Ferraz, Simple components and central units in group algebras. J. Algebra **279**(1), 191–203 (2004)
27. R.A. Ferraz, *Units of $\mathbb{Z}C_p$, Groups, Rings and Group Rings*, vol. 499, Contemporary Mathematics (American Mathematical Society, Providence, 2009), pp. 107–119
28. R.A. Ferraz, P.M. Kitani, Units of $\mathbb{Z}C_{p^n}$. Commun. Algebra **43**(11), 4936–4950 (2015)
29. R.A. Ferraz, R. Marcuz, Units of $\mathbb{Z}(C_p \times C_2)$ and $\mathbb{Z}(C_p \times C_2 \times C_2)$. Commun. Algebra **44**(2), 851–872 (2016)
30. R.A. Ferraz, J.J. Simón-Pınero, Central units in metacyclic integral group rings. Commun. Algebra **36**(10), 3708–3722 (2008)
31. R.A. Ferraz, J.J. Simón, Central units in $\mathbb{Z}C_p, q$. Commun. Algebra **44**(5), 2264–2275 (2016)
32. A. Giambruno, E. Jespers, Central idempotents and units in rational group algebras of alternating groups. Int. J. Algebra Comput. **8**(4), 467–477 (1998)
33. E.G. Goodaire, M.M. Parmenter, Units in alternative loop rings. Israel J. Math. **53**(2), 209–216 (1986)
34. A.W. Hales, I.B.S. Passi, Jordan decomposition, Algebra, Trends Math. (Birkhäuser, Basel, 1999), pp. 75–87
35. A.W. Hales, I.B.S. Passi, Group rings and Jordan decomposition. textitGroups, Rings, Group Rings, and Hopf Algebras, vol. 688, Contemporary Mathematics (American Mathematical Society, Providence, 2017), pp. 103–111
36. M. Hertweck, Contributions to the integral representation theory of groups. Habilitationsschrift (2004). http://elib.uni-stuttgart.de/handle/11682/4734

37. M. Hertweck, E. Jespers, Class-preserving automorphisms and the normalizer property for Blackburn groups. J. Group Theory **12**(1), 157–169 (2009)
38. M. Hertweck, E. Iwaki, E. Jespers, S.O. Juriaans, On hypercentral units in integral group rings. J. Group Theory **10**(4), 477–504 (2007)
39. G. Higman, The units of group-rings. Proc. London Math. Soc. (2) **46**, 231–248 (1940)
40. K. Hoechsmann, Units and class-groups in integral elementary abelian group rings. J. Pure Appl. Algebra **47**, 253–264 (1987)
41. K. Hoechsmann, Constructing units in commutative group rings. Manuscripta Math. **75**(1), 5–23 (1992)
42. K. Hoechsmann, S.K. Sehgal, A. Weiss, Cyclotomic units and the unit group of an elementary abelian group ring. Arch. Math. **45**(1), 5–7 (1985)
43. B. Huppert, *Endliche Gruppen. I*, Die Grundlehren der Mathematischen Wissenschaften, Band 134 (Springer, Berlin-New York, 1967)
44. E. Iwaki, *Unidades Hipercentrais em Anéis de Grupo Inteiro e a Hiperbolicidade do Grupo de Unidades de uma Álgebra de Grupo Modular*, Instituto de Matemática e Estatístíca de Universidade de São Paulo, Brazil (2006). https://doi.org/10.11606/D.45.2000.tde-20052007-112821
45. E. Iwaki, S.O. Juriaans, Hypercentral unit groups and the hyperbolicity of a modular group algebra. C. R. Math. Acad. Sci. Soc. R. Can. **29**(2), 61–64 (2007)
46. E. Iwaki, S.O. Juriaans, Hypercentral unit groups and the hyperbolicity of a modular group algebra. Commun. Algebra **36**(4), 1336–1345 (2008)
47. E. Jespers, Units in integral group rings: a survey, *Methods in Ring Theory*, vol. 198, Lecture Notes in Pure and Applied Mathematics (1997), pp. 141–169
48. E. Jespers, Á. del Río, *Group Ring Groups*, Volume 1: Orders and Generic Constructions of Units, De Gruyter, Berlin-Boston (2015)
49. E. Jespers, M.M. Parmenter, Construction of central units in integral group rings of finite groups. Proc. Amer. Math. Soc. **140**(1), 99–107 (2012)
50. E. Jespers, G. Olteanu, Á. del Río, I. Van Gelder, Central units of integral group rings. Proc. Amer. Math Soc. **142**, 2193–2209 (2014)
51. E. Jespers, M.M. Parmenter, S.K. Sehgal, Central units of integral group rings of nilpotent groups. Proc. Amer. Math. Soc. **124**(4), 1007–1012 (1996)
52. E. Jespers, S.O. Juriaans, J.M. de Miranda, J.R. Rogerio, On the normalizer problem. J. Algebra **247**(1), 24–36 (2002)
53. E. Jespers, G. Olteanu, Á. del Río, I. Van Gelder, Group rings of finite strongly monomial groups: central units and primitive idempotents. J. Algebra **387**, 99–116 (2013)
54. E. Jespers, Á. del Río, I. Van Gelder, Writing units of integral group rings of finite abelian groups as a product of Bass units. Math. Comp. **83**(285), 461–473 (2014)
55. Y. Li, The hypercentre and the n-centre of the unit group of an integral group ring. Canad. J. Math. **50**(2), 401–411 (1998)
56. Y. Li, M.M. Parmenter, Central units of the integral group ring $\mathbf{Z}A_5$. Proc. Amer. Math. Soc. **125**(1), 61–65 (1997)
57. Y. Li, M.M. Parmenter, Hypercentral units in integral group rings. Proc. Am. Math. Soc. **129**(8), 2235–2238 (2001)
58. Y. Li, M.M. Parmenter, Some results on hypercentral units in integral group rings. Commun. Algebra **31**(7), 3207–3217 (2003)
59. Y. Li, M.M. Parmenter, The upper central series of the unit group of an integral group ring. Commun. Algebra **33**(5), 1409–1415 (2005)
60. Y. Li, M.M. Parmenter, Central units in integral group rings II. Int. J. Algebra **8**(1–4), 47–55 (2014)
61. Y. Li, S.K. Sehgal, M.M. Parmenter, On the normalizer property for integral group rings. Commun. Algebra **27**(9), 4217–4223 (1999)
62. I.D. Macdonald, *Some p-groups of Frobenius and extra-special type*. Israel J. Math. **40** (1981), no. 3-4, 350–364 (1982)

63. S. Maheshwary, Integral group rings with all central units trivial: solvable groups. Indian J. Pure Appl. Math. **49**(1), 169–175 (2018)
64. A. Olivieri, Á. del Río, J.J. Simón, On monomial characters and central idempotents of rational group algebras. Commun. Algebra **32**(4), 1531–1550 (2004)
65. M.M. Parmenter, in *Algebra: Some Recent Advances*, ed. by I.B.S. Passi. Central Units in Integral Group Rings, Trends Math., Birkhäuser, Basel, 111–116 (1999)
66. C. Polcino Milies, S.K. Sehgal, Central units of integral group rings. Commun. Algebra **27**(12), 6233–6241 (1999)
67. C. Polcino Milies, S.K. Sehgal, *An Introduction to Group Rings*, Algebras and Applications, vol. 1 (Kluwer Academic Publishers, Dordrecht, 2002)
68. J. Ritter, S.K. Sehgal, Integral group rings with trivial central units. Proc. Amer. Math. Soc. **108**(2), 327–329 (1990)
69. J. Ritter, S.K. Sehgal, Units of group rings of solvable and Frobenius groups over large rings of cyclotomic integers. J. Algebra **158**(1), 116–129 (1993)
70. J. Ritter, S.K. Sehgal, Trivial units in RG. Math. Proc. R. Ir. Acad. **105A**(1), 25–39 (2005)
71. S.K. Sehgal, Units in commutative integral group rings. Math. J. Okayama Univ. **14**, 135–138 (1969/1970)
72. S.K. Sehgal, *Topics in Group Rings*, vol. 50, Monographs and Textbooks in Pure and Applied Math (Marcel Dekker Inc, New York, 1978)
73. S.K. Sehgal, *Units in Integral Group Rings*, vol. 69, Pitman Monographs and Surveys in Pure and Applied Mathematics (Longman Scientific & Technical, Harlow, 1993), With an appendix by Al Weiss
74. S.K. Sehgal, H.J. Zassenhaus, Integral group rings with nilpotent unit groups. Commun. Algebra **5**(2), 101–111 (1977)
75. The GAP Group, *GAP – Groups, Algorithms, and Programming, Version 4.5.6* (2012)
76. L.C. Washington, *Introduction to Cyclotomic Fields*, 2nd ed., , vol. 83, Graduate Texts in Mathematics (Springer, New York, 1997)

Character Tables and Sylow Subgroups Revisited

Gabriel Navarro

2010 Mathematics Subject Classification Primary 20C15

1 Introduction

It is more than 14 years since I wrote in [22] a list of open problems on characters and Sylow subgroups for the Proceedings of the John Thompson conference (Gainesville, Florida, 2003). In this lapse of time, there has been significant progress on some of these questions. Others have remained unaccessible or simply ignored. It is my purpose here to give an account on what has been done in these years, add a few more problems, and comment on others.

It is hardly debatable that one of the corner stones of group theory is Sylow theory. For a character theorist with love for finite groups, there are few, if any, more interesting subjects than the study of the relationship between the set $\mathrm{Irr}(G)$ of the irreducible complex characters of G and $P \in \mathrm{Syl}_p(G)$, a Sylow p-subgroup of G. If one wants to go deeper, there is a more sophisticated version of this: relate the irreducible characters in a Brauer p-block $\mathrm{Irr}(B)$ with D, where D is a defect group of B.

Research supported by the Prometeo/Generalitat Valenciana, Proyecto MTM2016-76196-P and FEDER funds.

G. Navarro (✉)
Departament of Mathematics, Universitat de València, Burjassot 46100 Valencia, Spain
e-mail: gabriel.navarro@uv.es

N. S. N. Sastry and M. K. Yadav (eds.), *Group Theory and Computation*,
Indian Statistical Institute Series, https://doi.org/10.1007/978-981-13-2047-7_10

2 Brauer' Problems

As many of the fundamental questions in the Representation Theory of Finite Groups, it all starts with Richard Brauer. There are some problems in his celebrated paper [3] that have totally shaped the research in our subject. This is Brauer's Problem 12, in Brauer's words.

Problem 12. *Given the character table of a group G and a prime p dividing $n = |G|$, how much information about the structure of the p-Sylow group P can be obtained? In particular, can it be decided whether or not P is Abelian?*

W. Kimmerle and R. Sandling solved this problem (the abelian part) in 1995 [18], but perhaps not in the way that Brauer was thinking. Kimmerle and Sandling showed, using the classification of finite simple groups, that if G and H have the same character table, then G has abelian Sylow p-subgroups if and only if H has abelian Sylow p-subgroups. (They even proved another remarkable fact: that those Sylow subgroups have to be isomorphic.) But their method is not practical. If we have a character table, how do we recognize if the Sylow p-subgroup is abelian?

Of course, Brauer was thinking about his famous Height Zero Conjecture on blocks (Problem 23 in [3]) when he proposed Problem 12.

There are many ways to introduce blocks, but from the character theory point of view, the best way is to use something that can be computed in the character table. It is time to remark that the character table $X(G)$ of a finite group G is the square complex matrix $(\chi_i(x_j))$ whose (i, j)-entry is the value of the irreducible character χ_i on the representative x_j of the conjugacy classes of G. (Of course, this is only well defined up to a permutation of rows and columns.) We do not know the orders of the elements x_j, although we do know the set of primes dividing $o(x_j)$. (This is a theorem of G. Higman, see Theorem 8.21 of [11].) A way to define p-blocks is this: we define a linking $\leftrightarrow$ in the set $\mathrm{Irr}(G)$ by linking $\alpha \leftrightarrow \beta$ if and only if

$$\sum_{x \in G^0} \alpha(x)\overline{\beta(x)} \neq 0\,,$$

where G^0 is the set of p-regular elements of G. Then the **blocks** are the connected components in $\mathrm{Irr}(G)$ via this linking. If $B \subseteq \mathrm{Irr}(G)$ is a block, then the characters χ in B with smallest possible $\chi(1)_p$ are called the **height zero** characters of the block, and it turns out that if $\alpha \in B$ has height zero, then $\chi \in \mathrm{Irr}(G)$ belongs to B if and only if

$$\sum_{x \in G^0} \alpha(x)\overline{\chi(x)} \neq 0\,.$$

(This is Corollary 3.25 of [20].) The **principal block** of G is the block containing the trivial character, and therefore the height zero characters in this block are the characters of degree not divisible by p.

Conjecture 2.1 (Brauer's Height Zero, for principal blocks) *Let G be a finite group, and let $P \in \mathrm{Syl}_p(G)$. Then, P is abelian if and only if whenever $\sum_{x \in G^0} \chi(x) \neq 0$, where $\chi \in \mathrm{Irr}(G)$, then p does not divide $\chi(1)$.*

We see that Conjecture 2.1 gives an explicit algorithm to compute from the character table $X(G)$ if $P \in \mathrm{Syl}_p(G)$ is abelian. Since the writing of [22], there has been fundamental progress on Brauer's Height Zero Conjecture (for every block). In [17], one direction of the conjecture was proven (if the *defect group* of B is abelian, then all irreducible characters in B have height zero). In [28], it was proven for 2 blocks of maximal defect, and the connection with the so called Alperin–McKay conjecture was foreseen. Later in [30], with the help of [29], the remaining direction was proven to be a consequence of the *Inductive Alperin-McKay condition* [41], and therefore, it was reduced to a question on simple groups.

But in fact, in the same paper [3], Brauer was also wondering about an easier way to detect if P is abelian from the character table of G: *"If we know the p-classes K, we have a necessary condition for this question: $c(K)$ for these classes must be divisible by the full power of p dividing n. It does not seem to be known whether this condition is sufficient for P to be Abelian."* (Here, $c(K)$ is the size of the centralizer of any element in K.)

Brauer, of course, was not aware of the examples J_4 or Ru for $p = 3$, and Th for $p = 5$. A. Camina and M. Herzog [4] proved that Brauer was indeed right for $p = 2$. (Much later E. Henke gave another proof of this result in [10].) Somewhat surprisingly, Tiep and this author proved in [31] that the cases $p = 3$ and $p = 5$ were the only exceptions.

Theorem 2.2 *Suppose that $p \neq 3, 5$. Let G be a finite group, and $P \in \mathrm{Syl}_p(G)$. Then, P is abelian if and only if p does not divide the sizes of the conjugacy classes of the p-elements of G.*

Afterward, with R. Solomon, we completely characterized the finite groups satisfying the condition in Theorem 2.2 [37], and were able to give a complete solution to Brauer's Problem 12. (Of course, while we wait for the entire resolution of the Height Zero Conjecture).

Theorem 2.3 *Let G be a finite group, let p be any prime and $P \in \mathrm{Syl}_p(G)$. Then, P is abelian if and only if every p-element has conjugacy class size not divisible by p, and, if $p = 3$ or 5, whenever $\sum_{x \in G^0} \chi(x) \neq 0$, where $\chi \in \mathrm{Irr}(G)$, then p does not divide $\chi(1)$.*

Let me digress for a moment and change the subject of abelian Sylow to nilpotent Hall subgroups and character tables. (This is related to Brauer's Problem 11: *Given the character table of a group G, how much information about the existence of subgroups can be obtained?*) Again, Kimmerle and Sandling proved that if $X(G) = X(H)$ and G has a nilpotent Hall π-subgroup, then H has a nilpotent Hall π-subgroup. But, in [2], we found a natural condition that easily characterizes this fact in the character table.

Theorem 2.4 *Let G be a finite group. Then, G has nilpotent Hall subgroups if and only if for every pair of distinct primes $p, q \in \pi$ the class sizes of the p-elements of G are not divisible by q.*

Besides of the questions that I formulated in [22] for which there has been no progress or activity, we believe that it might be interesting to work on the following.

Problems 2.5 *(a) Does the multi-set of character degrees of a finite group G determine whether G has abelian Sylow p-subgroups?*

(b) Does the character table of a finite group G determine whether G possesses Hall π-subgroups? Or even whether G possesses Hall π-subgroups H satisfying that every π-subgroup of G is contained in a G-conjugate of H?

(c) Does the character table $X(G)$ determine, for instance, whether $P \in \mathrm{Syl}_p(G)$ has class 2?

Problem 2.5(a) is asking if only the first column of the character table of G determines if G has abelian Sylow p-subgroups! This is highly unlikely, but at the time of this writing I have no counterexample to this. For Problem 2.5(b), notice that if H is a Hall π-subgroup H containing up to G-conjugacy the π-subgroups of G, then we have a permutation character $(1_H)^G$ with many nice properties. Whether or not the properties of such a character determines the existence of H seems like an interesting problem.

3 The McKay Conjecture

As we all know, the McKay conjecture is at the center of the Representation Theory of Finite Groups today. It asserts that if G is a finite group and $P \in \mathrm{Syl}_p(G)$, then

$$|\mathrm{Irr}_{p'}(G)| = |\mathrm{Irr}_{p'}(\mathbf{N}_G(P))|,$$

where $\mathrm{Irr}_{p'}(G)$ is set of the irreducible characters of G of degree not divisible by p. In 2007 [15], we reduced the McKay conjecture to a problem on simple groups: if the finite simple groups satisfy what is now called the *inductive McKay condition*, then the McKay conjecture is true. This approach has proven to be successful, as shown in the landmark paper by G. Malle and B. Späth [19], in which they prove the McKay conjecture for $p = 2$.

But there is a strong conviction that there has to be more than "merely" bijections

$$^* : \mathrm{Irr}_{p'}(G) \to \mathrm{Irr}_{p'}(\mathbf{N}_G(P)).$$

For instance, we believe that there should be bijections * satisfying $\chi(1) \equiv \pm\chi^*(1) \mod p$ [13]; such that

$$(\chi^*)^a = (\chi^a)^*$$

for all $a \in \mathrm{Aut}(G)$ that stabilize P [24]; that these bijections should respect *cohomology* [15]; Brauer blocks and the Brauer correspondence [1]; and, what concerns us here, that they should commute with certain Galois action, as proposed in [23].

Our aim in the rest of this paper is to survey how the McKay–Galois conjecture is giving us insight on the problem of relating character tables and local structure.

By elementary character theory, we have that $\mathrm{Irr}_{p'}(\mathbf{N}_G(P)) = \mathrm{Irr}(\mathbf{N}_G(P)/P')$, where P' is the derived subgroup of P. The group $\mathbf{N}_G(P)/P'$ is the semidirect product of P/P' with $\mathbf{N}_G(P)/P$ and in particular, the groups P/P' and $\mathbf{N}_G(P)/P$ are natural targets of research.

4 The Group $\mathbf{N}_G(P)/P$

The most basic question here is: Does the character table detect if this group is trivial? The McKay–Galois conjecture provides an answer to this. And for odd primes p, this has become a theorem (not a conjecture). Recall that an irreducible character χ is **p-rational** if the field of values $\mathbb{Q}(\chi)$ is contained in some cyclotomic field $\mathbb{Q}_n$, for n not divisible by p. Of course, if G is a p'-group, then every irreducible character of G is p-rational. Since the McKay–Galois conjecture (which we have not stated yet) implies that the number of p-rational characters in $\mathrm{Irr}_{p'}(G)$ and $\mathrm{Irr}_{p'}(\mathbf{N}_G(P))$ is the same, it easily follows from this fact that, for odd primes, we have that $\mathbf{N}_G(P) = P$ if and only if the unique p'-degree p-rational irreducible character of G is the trivial one. This result was proven in [35].

Theorem 4.1 *Let p be an odd prime, let G be a finite group, and let $P \in \mathrm{Syl}_p(G)$. Then $\mathbf{N}_G(P) = P$ if and only if the trivial character is the only p-rational p'-degree irreducible character of G.*

Our interest in self-normalizing Sylow subgroups started when we tried to prove the McKay conjecture in this case. The first fact that we discovered was that if $\mathbf{N}_G(P) = P$, then G is solvable for $p > 3$ [9]. Almost at the same time [21], I proved that for solvable groups G with $\mathbf{N}_G(P) = P$, there is a natural bijection

$$^* : \mathrm{Irr}_{p'}(G) \to \mathrm{Irr}(P/P') .$$

This bijection can be described quite easily: If $\chi \in \mathrm{Irr}_{p'}(G)$, then

$$\chi_P = \chi^* + \Delta ,$$

where $\chi^* \in \mathrm{Irr}(P)$ is linear and every irreducible constituent of Δ (if any) has degree divisible by p. Later in [36], we proved the same result for every finite group, for p odd. In particular, since * commutes with Galois action, notice that Theorem 4.1 can easily be proved to be a consequence of this correspondence.

The subject of when there exist natural McKay bijections is quite interesting (see [6, 8, 16]). It is also a subtle subject, since the word *natural* has never been defined in character theory. We understand that a map is canonical, natural, or choice-free when the outcome does not depend on any choice made in order to define it. Perhaps, there is not even need for a definition: one simply recognizes a canonical map when one sees it. In fact, we digress, one of the reasons why the McKay conjecture is so hard might be because there are not natural bijections between $\mathrm{Irr}_{p'}(G)$ and $\mathrm{Irr}_{p'}(\mathbf{N}_G(P))$, in general. This might be even related to the title of the book of the American psychologist Barry Schwartz: "The paradox of choice: why more is less".

Inspired by the block version of the McKay–Galois conjecture, we can detect further local structure in the character table. The following is the main result of [27] and lies much deeper than Theorem 4.1.

Theorem 4.2 *Let G be a finite group, let p be an odd prime, and let $P \in \mathrm{Syl}_p(G)$. Then, the trivial character is the only p'-degree p-rational character in the principal block of G if and only if $\mathbf{N}_G(P) = P \times K$.*

After Theorems 4.1 and 4.2, the question left is what happens with all these results for $p = 2$. At the time of this writing, we have no theorems but conjectures.

A single Galois automorphism holds the key in this situation. Let σ be the Galois automorphism that fixes two power roots of unity and squares odd roots of unity.

Conjecture 4.3 *Let G be a finite group, and let $P \in \mathrm{Syl}_2(G)$. Then, $P = \mathbf{N}_G(P)$ if and only if every irreducible odd-degree character of G is σ-invariant.*

This conjecture was reduced to almost simple groups in [40]. The block version of it, which can be proved to imply Conjecture 4.3, has also been reduced to almost simple groups in [34].

Conjecture 4.4 *Let G be a finite group, and let $P \in \mathrm{Syl}_2(G)$. Then, $\mathbf{N}_G(P) = P \times K$ if and only if every odd-degree irreducible character in the principal block of G is σ-invariant.*

Again, all these conjectures are consequences of the McKay–Galois conjecture. Recent work of A. Schaeffer-Fry and J. Taylor makes us believe that they might turn into theorems soon.

Question 7 in [22] asked if $X(G)$ determines $|\mathbf{N}_G(P)|$. Since $X(G)$ determines $|P| = |G|_p$, this is a more general question than asking if $\mathbf{N}_G(P) = P$. If G is p-solvable, there has been progress in [33]. Recall that the p-power map of a character table is the function f defined by the following: If $\{x_1, \ldots, x_k\}$ are representatives of the conjugacy classes of G then $f : \{1, \ldots, k\} \to \{1, \ldots, k\}$ is defined such that $(x_j)^p$ lies in the class of $x_{f(j)}$.

Theorem 4.5 *Suppose that G is p-solvable, and let $P \in \mathrm{Syl}_p(G)$. Then, $X(G)$ and the p-power map determine $|\mathbf{N}_G(P)|$.*

Besides of Question 9 that I formulated in [22] (to decide if $X(G)$ determines if $\mathbf{N}_G(P)/P$ is abelian) for which there has been no progress, it might be interesting to work on the following.

Problems 4.6 *(a) How much does $X(G)$ know about the structure of* $\mathbf{N}_G(P)/P$*? Does it determine the set of its prime divisors? Or if* $\mathbf{N}_G(P)/P$ *is cyclic?*

(b) Is there a canonical subset $\mathcal{A}$ of $\mathrm{Irr}_{p'}(G)$ *which we can naturally associate to the set* $\mathrm{Irr}(\mathbf{N}_G(P)/P)$*? For instance, if G is p-solvable, then $\mathcal{A} = \mathcal{X}_{p'}(G)$ is the set of Gajendragadkar p'-special characters of G (See [5]). Or if P is cyclic, then $\mathcal{A}$ is the set of the non-exceptional characters of G, as defined in the cyclic defect theory.*

E. Giannelli, in private communication, has constructed $\mathcal{A}$ for symmetric groups [7]. It is not clear what $\mathcal{A}$ could be for groups of Lie type. We believe that $\mathcal{A}$ should at least consist of p-rational characters, and their degrees should be congruent with plus or minus the degrees of the irreducible characters in $\mathbf{N}_G(P)/P$.

5 The Group P/P'

In Sect. 2, we essentially wrote on the case where P is abelian, that is, when $P' = 1$. The main general question in this section is how much $X(G)$ knows about the group P/P'. For instance, Question 1 of [22] asked if $|P/P'|$ is determined by $X(G)$. We do not know if this is true, even for p-solvable groups.

Again, the McKay–Galois conjecture provides a relationship between $X(G)$ and P/P'. If one thinks about it, it is surprising that a tiny local section of a finite group influences (and is influenced by) its character table. The following is one of the origins of what later became the Galois–McKay conjecture [12].

Conjecture 5.1 *Let $e \geq 1$. Let σ be the Galois automorphism of* $\mathrm{Gal}(\mathbb{Q}^{\mathrm{ab}})$ *that fixes roots of unity of order not divisible by p, and sends p-power roots of unity ξ to ξ^{1+p^e}. Let G be a finite group, and let $P \in \mathrm{Syl}_p(G)$. Then all the irreducible characters of p'-degree of G are σ-fixed if and only if the exponent of P/P' is less than or equal to p^e.*

There is quite a recent development in this conjecture [26].

Theorem 5.2 *Let G be a finite group, and let $P \in \mathrm{Syl}_p(G)$. If all the irreducible characters of p'-degree in the principal block of G are σ-fixed, then the exponent of P/P' is less than or equal to p^e. The converse holds if it holds for central extensions of almost simple groups.*

The exponent of P/P' for $P \in \mathrm{Syl}_2(G)$ has also received attention in a conjecture of R. Gow that has been recently proved in [32]. Unfortunately, it only goes in one direction.

Theorem 5.3 *Let G be a finite group, and let $P \in \mathrm{Syl}_2(G)$. If all the irreducible characters of odd-degree in the principal block of G are real valued, then P/P' is elementary abelian.*

Theorem 5.3 might be suggesting something that it is simply not true. It is false that if all characters in $\mathrm{Irr}(G)$ are real valued, then all characters in $\mathrm{Irr}(P)$ are real valued, and there are some easy examples. The same statement is also false for rational-valued characters, but this was a 50-year-old conjecture which was disproved in [14].

It is perhaps time to state the McKay–Galois conjecture. Let $\mathcal{H} \subseteq \mathrm{Gal}(\bar{\mathbb{Q}}/\mathbb{Q})$ be the group of all field automorphisms σ that send each p'-root of unity ξ to some p-power ξ^{p^e}, where e is a fixed but arbitrary nonnegative integer depending on σ.

Conjecture 5.4 *The actions of $\mathcal{H}$ on $\mathrm{Irr}_{p'}(G)$ and $\mathrm{Irr}_{p'}(\mathbf{N}_G(P))$ are permutation isomorphic.*

When trying to find new connections between $\mathrm{Irr}_{p'}(G)$ and $\mathrm{Irr}_{p'}(\mathbf{N}_G(P))$ in the McKay conjecture, it is easy to see that there cannot always exist McKay bijections that commute with the action of the whole group $\mathrm{Gal}(\bar{\mathbb{Q}}/\mathbb{Q})$. This would imply, for instance, that the number of rational-valued (or real) characters in $\mathrm{Irr}_{p'}(G)$ and $\mathrm{Irr}_{p'}(\mathbf{N}_G(P))$ is the same, and this is plainly false. There are plenty of counterexamples to this assertion. (For instance, $G = \mathrm{GL}_2(3)$ for $p = 3$.) If one thinks about it, this is too much to expect, since we are holding fixed a prime p in the McKay problem and we are not taking into account the prime p at this point. The subgroup of $\mathrm{Gal}(\bar{\mathbb{Q}}/\mathbb{Q})$ that takes p naturally into account is precisely $\mathcal{H}$, because it is exactly the stabilizer of the prime ideals containing p in the ring of algebraic integers of the cyclotomic field $\mathbb{Q}_n$ for every n.

If the McKay conjecture forces us to understand how $\mathrm{Aut}(S)$ acts on $\mathrm{Irr}(S)$ for every simple group S (an open and extremely hard problem on its own), the McKay–Galois forces us not only to understand this action, but to control Galois actions, and therefore, character values. Contrary to the case of the ordinary McKay conjecture, there is no reduction of the Galois version of it, although some attempts are under way. It seems that Turull's Brauer–Clifford group [42] might be important in that.

(NOTE. There has been some recent developments since the writing of this paper. B. Sambale has found a counterexample to Problem 2.5(a) for $p = 2$ in [25]. A. Schaeffer-Fry has proven Conjecture 4.3 in [38, 39]. E. Giannelli [7] has found a canonical set for the symmetric groups S_n for Problem 4.6(b). There is now a reduction of the McKay-Galois conjecture by B. Späth, C. Vallejo and myself.)

References

1. J.L. Alperin, Local representation theory. Proc. Symp. Pure Math. **37**, 369–375 (1980)
2. A. Beltrán, M.J. Felipe, G. Malle, A. Moretó, G. Navarro, L. Sanus, R. Solomon, P.H. Tiep, Nilpotent and Abelian hall subgroups in finite groups. Trans. AMS. **368**, 2497–2513 (2016)
3. R. Brauer, Representations of finite groups, in *Lectures on Modern Mathematics*, vol. I, ed. by T. Saaty (Wiley, 1963)
4. A. Camina, M. Herzog, Character tables determine abelian Sylow 2-subgroups. Proc. Amer. Math. Soc. **80**, 533–535 (1980)

5. D. Gajendragadkar, A characteristic class of characters of finite π-separable groups. J. Algebra **59**, 237–259 (1979)
6. E. Giannelli, Characters of odd degree of symmetric groups. https://arxiv.org/pdf/1601.03839.pdf
7. E. Giannelli, Private communication
8. E. Giannelli, A. Kleschev, G. Navarro, P.H. Tiep, Restriction of odd degree characters and natural correspondences. Int. Math. Res. Not. No. 00, 1–30 (2016)
9. R.M. Guralnick, G. Malle, G. Navarro, Self-normalizing Sylow subgroups. Proc. Amer. Math. Soc. **132**(4), 973–979 (2004)
10. E. Henke, A characterization of saturated fusion systems over abelian 2-groups. Adv. Math. **257**, 1–5 (2014)
11. I.M. Isaacs, *Character Theory of Finite Groups* (Academic Press, 1976)
12. I.M. Isaacs, G. Navarro, Characters of p'-degree of p-solvable groups. J. Algebra **246**(1), 394–413 (2001)
13. I.M. Isaacs, G. Navarro, New refinements of the McKay conjecture for arbitrary finite groups. Ann. of Math. (2) **156**(1), 333–344 (2002)
14. I.M. Isaacs, G. Navarro, Sylow 2-subgroups of rational solvable groups. Math. Z. **272**(3–4), 937–945 (2012)
15. I.M. Isaacs, G. Malle, G. Navarro, A reduction theorem for the McKay conjecture. Invent. Math. **170**, 33–101 (2007)
16. I.M. Isaacs, G. Navarro, J. Olsson, P.H. Tiep, Character restrictions and multiplicities in symmetric groups. J. Algebra **478**, 271–282 (2017)
17. R. Kessar, G. Malle, Quasi-isolated blocks and Brauer's height zero conjecture. Ann. Math. **178**, 321–384 (2013)
18. W. Kimmerle, R. Sandling, Group theoretic determination of certain Sylow and Hall subgroups and the resolution of a question of R. Brauer. J. Algebra **171**, 329–346 (1995)
19. G. Malle, B. Späth, Characters of odd degree. Ann. of Math. (2)**184**(3), 869–908 (2016)
20. G. Navarro, *Characters and Blocks of Finite Groups*, London Mathematical Society Lecture Note Series 250 (Cambridge University Press, Cambridge, 1998)
21. G. Navarro, Linear characters of Sylow subgroups. J. Algebra **269**(2), 589–598 (2003)
22. G. Navarro, Problems on characters and Sylow subgroups. Finite groups 2003, 275–281 (Walter de Gruyter GmbH Co. KG, Berlin, 2004)
23. G. Navarro, The McKay conjecture and Galois automorphisms. Ann. Math. **160**, 1129–1140 (2004)
24. G. Navarro, Actions and characters in blocks. J. Algebra **275**, 471–480 (2004)
25. G. Navarro, B. Sambale, On the blockwise isomorphism problem. Manuscripta Mathematica (to appear)
26. G. Navarro, P.H. Tiep, Sylow subgroups, exponents and character tables, in preparation
27. G. Navarro, P.H. Tiep, C. Vallejo, Local blocks with one simple module, submitted
28. G. Navarro, P.H. Tiep, Brauer's height zero conjecture for the 2-blocks of maximal defect. J. Reine Angew. Math. **669**, 225–247 (2012)
29. G. Navarro, P.H. Tiep, Characters of relative p'-degree with respect to a normal subgroup. Ann. Math. **178**(3), 1135–1171 (2013)
30. G. Navarro, B. Späth, On Brauer's height zero conjecture. J. Eur. Math. Soc. **16**, 695–747 (2014)
31. G. Navarro, P.H. Tiep, Abelian Sylow subgroups in a finite group. J. Algebra **398**, 519–526 (2014)
32. G. Navarro, P.H. Tiep, Real groups and Sylow 2-subgroups. Adv. Math. **299**, 331–360 (2016)
33. G. Navarro, N. Rizo, A Brauer-Wielandt formula (with an application to character tables). Proc. Amer. Math. Soc. **144**(10), 4199–4204 (2016)
34. G. Navarro, C. Vallejo, 2-Local blocks with one simple module. J. Algebra **488**, 230–243 (2017)
35. G. Navarro, P.H. Tiep, A. Turull, p-rational characters and self-normalizing Sylow p-subgroups. Represent. Theory **11**, 84–94 (2007)

36. G. Navarro, P.H. Tiep, C. Vallejo, McKay natural correspondences of characters. Algebra Number Theory **8**, 1839–1856 (2014)
37. G. Navarro, R. Solomon, P.H. Tiep, Abelian Sylow subgroups of finite groups II. J. Algebra **421**, 3–11 (2015)
38. A. Schaeffer-Fry, J. Taylor, On Self-Normalising Sylow 2-Subgroups in Type A. https://arxiv.org/pdf/1701.00272.pdf
39. A. Schaeffer-Fry. https://arxiv.org/pdf/1707.03923.pdf
40. A. Schaeffer-Fry, Odd-degree characters and self-normalizing Sylow 2-subgroups: a reduction to simple groups. Commun. Algebra **44**, 1882–1905 (2016)
41. B. Späth, A reduction theorem for the Alperin-McKay conjecture. J. reine angew. Math. **680**, 153–189 (2013)
42. A. Turull, The Brauer-Clifford group. J. Algebra **321**, 3620–3642 (2009)

Printed in the United States
By Bookmasters